**A Natural History Guide to
Great Smoky Mountains
National Park**

A Natural History Guide to Great Smoky Mountains National Park

Donald W. Linzey

The University of Tennessee Press / Knoxville

Copyright © 2008 by The University of Tennessee Press / Knoxville.
All Rights Reserved. Printed in China.
First Edition.

Unless otherwise credited, photographs are courtesy of Great Smoky Mountains National Park.

This book is printed on acid-free paper.

Library of Congress Cataloging-in-Publication Data

Linzey, Donald W.
  A natural history guide to Great Smoky Mountains National Park / Donald W. Linzey. — 1st ed.
   p. cm.
Includes bibliographical references and index.
ISBN-13: 978-1-57233-612-4 (pbk. : alk. paper)
ISBN-10: 1-57233-612-9

1. Natural history—Great Smoky Mountains National Park (N.C. and Tenn.)
2. Great Smoky Mountains National Park (N.C. and Tenn.)
I. Title.
  QH105.N8L56 2008
  508.768'89—dc22   2007047695

To Mom and Dad,
who introduced me to the Smokies;
To my sister, June,
who was with me on my first two trips;
and
To my loving and supportive wife, Nita,
with whom I have enjoyed the Smokies
for the past 24 years.

I don't ask for the meaning of the song of a bird or the rising of the sun on a misty morning. There they are, and they are beautiful.

—Pete Hamill

Contents

Foreword	xv
Donald J. Shure	
Acknowledgments	xvii
Introduction	1
1. The Beginning	5
2. Topography and Climate	21
3. Pre-Park History	31
4. Park Formation	41
5. Forests and Balds	51
6. Animals	97
7. Endangered Species	161
8. Reintroductions	171
9. Natural History Research in the Park	183
10. All Taxa Biodiversity Inventory	197
11. Environmental Concerns	207
12. What the Future May Hold	235
Appendix I. Where Is That? Park Localities Referenced in Text	245
Appendix II. Checklist of Trees, Shrubs, and Woody Vines Referenced in Text	253
Appendix III. Checklist of Wildflowers, Herbs, Sedges, and Grasses Referenced in Text	257
Appendix IV. Checklist of Ferns Referenced in Text	261
Notes	263
Selected Bibliography	277
Index	279

Illustrations

Figures and Maps

Park Entrance Sign at Gatlinburg Entrance. xx
Great Smoky Mountains National Park (Map) 3
Movement of the Continents (Map). 10
Bedrock Geology of the Park (Map). 12
Baskins Creek Falls Showing Thunderhead Sandstone Formation. 15
Chimney Tops in Winter 15
View of Cades Cove from Gregorys Bald 17
Gregory Cave. 18
Slide Area on Mount LeConte 19
Little River at Metcalf Bottoms 22
Showy Orchis 25
Yellow Lady's Slipper 25
Pink Lady's Slipper. 25
Deptford Pink 25
Autumn Colors in Sugarland Valley. 26
Winter along Cades Cove Loop Road. 28
Jim Shelton's Family in Front of American Chestnut Tree 38
Clearcutting near Elkmont 39
Elkmont 39
Disturbance History in the Park 40
Champion Fibre Mill at Smokemont 42
The Walker Sisters 43
Louisa Walker 44
Steve Woody's Place in Cataloochee Valley. 46

National Park Service Mandate 47
Plaque at Newfound Gap Commemorating the Park's Dedication 47
President Roosevelt Speaking at the Park Dedication. 48
Spatial and Elevational Relationships of Major Forest Types. 52
Yellow-Fringed Orchid 53
Flowering Dogwood. 53
Fire Pink. 54
Turkey Tail Fungus 54
Caesar's Amanita Mushroom 55
Distribution of Glaciers during the Ice Age (Map). 56
"Graybacks" near Chimneys Picnic Area 57
Decaying Logs 58
Vegetation in the Park (Map) 60
Spruce-Fir Forest 61
Fraser Fir. 62
Fraser Fir and Red Spruce Needles. 62
Spores on Leaflets of the Polypody Fern 63
American Woodsorrel 63
Painted Trillium 64
Pink Turtlehead. 64
Turk's Cap Lily. 64
Spruce-Fir Forest Seen from Clingmans Dome Road, 1979. 65
Effects of the Balsam Woolly Adelgid, 2001 66
Three-Lobed Leaves of the Striped or "Goosefoot" Maple 66
Striped Maple with Snail. 67
Blossom of the Catawba Rhododendron. 67
Hiker in a Bed of Fringed Phacelia near Indian Gap 68
Fringed Phacelia (Close-up) 68
Dog-Hobble. 68
British Soldier Lichen 69
Tuliptree Blossoms. 69
Cove Hardwood Forest. 71
Umbrella Magnolia 72
Flowers of Witch Hazel. 72
Yellow Birch on Stiltlike Roots 73
Evergreen Leaves of Rhododendron. 74

Trout Lily	75
Indian Pipe	75
Doll's Eyes	76
Jack-in-the-Pulpit	76
Walking Fern	77
Bird's Nest Fungus	78
Trailing Arbutus	79
Christmas Fern	80
"Fiddleheads"	81
Partridgeberry	82
Mountain Laurel	89
Sandmyrtle	90
Grass Bald with Grazing Sheep	92
Flame Azalea	93
Flame Azalea (Close-up)	93
Andrews Bald	94
Spence Field	95
Centipede	98
Millipede	98
Millipede	98
Land Snail	99
Walkingstick	99
Monarch Butterfly	100
Becky Nichols and Ted Grannan Catching Aquatic Insects	104
Electroshocking Fish in a Park Stream	108
Brook Trout	109
Rainbow Trout	109
Marbled Salamander	112
Spotted Salamander	112
Hellbender	112
Long-Tailed Salamander	113
Pigmy Salamander	113
Hen Wallow Falls near Cosby	114
Dusky Salamander	114
Red-Cheeked Salamander	116
American Toad	117

Green Frog	118
Wood Frog	120
Eastern Box Turtle	121
Five-Lined Skink	122
Anole	123
Eastern Garter Snake	125
Lateral View of Python Skull	126
Corn Snake	126
Northern Water Snake	127
Copperhead	128
Timber Rattlesnake	128
Heat-Sensing Facial Pits of Copperhead	128
Indigo Bunting	131
Pileated Woodpecker	133
Belted Kingfisher	136
Brown-Headed Cowbird	137
Wild Turkey	138
White-Breasted Nuthatch	139
Ruby-Throated Hummingbird	139
Figure-Eight Pattern of Hummingbird Beating Its Wings	139
Gray Fox	142
Opossum	142
Newborn Opossums	142
Spotted Skunk	143
Mink	144
Raccoon	144
Chipmunk	145
Red Squirrel	145
Coyote	146
White-Tailed Deer	148
White-Tailed Deer Fawn with Spots	149
Woodchuck	150
Short-Tailed Shrew	153
Mole	153
Bear Raiding Garbage Can	154
Bear Being Fed by Park Visitors	155

Bear #75	157
Rafinesque's Big-Eared Bat	159
Spruce-Fir Moss Spider	163
Red-Cockaded Woodpecker	164
Northern Flying Squirrel	166
Flying Squirrel Launching Itself	166
Eastern Cougar	168
Bobcat	169
Smoky Madtom	172
Abrams Falls	173
Biologists Reintroducing the Smoky Madtom to Abrams Creek	173
Duskytail Darter	175
River Otter	176
Elk	178
Red Wolf	180
Peregrine Falcon	181
Barn Owl	182
Art Stupka in 1943 with Visitors to the Park	184
Art Stupka in 1960	189
Don De Foe	191
Green Alga *Draparnaldia Appalachiana*	201
Subaerial Diatom *Decussata Placenta*	201
Tardigrade *Echiniscus Virginicus*	202
Air Quality Monitoring Station at Look Rock	209
Blackberry Leaves Showing Ozone Damage	210
Example of Southern Pine Beetle Damage	213
Balsam Woolly Adelgid Infestation	214
American Chestnut in Bloom	217
American Chestnut Blossom (Close-up)	217
American Chestnut Burr	218
Chestnut Blight	218
Dogwood Leaves Affected by Fungus	220
Male Gypsy Moth	222
Gypsy Moth Eggs on Tire	223
Gypsy Moth Eggs and Pupa	223
Gypsy Moth Larvae Eating Leaves	224

Hemlock Woolly Adelgid 225
Forest Technician Injecting Insecticide into Hemlock Trees 228
European Wild Hog 231
Typical Hog Damage on a Grass Bald 232

Tables

1.1. Geological Time Scale 9
1.2. Rock Types Found in Park 16
2.1. Top Ten Peaks in Park 23
2.2. Average Monthly Temperature and Precipitation Data for
     Gatlinburg, TN, and Clingmans Dome 24
4.1. Acreage of Park, 1940–2007 49
5.1. Twenty Tallest and Biggest Species of Trees in Park 83
5.2. 2006–2007 National Champion and Co-Champion Trees in Park 85
5.3. Acreage of Prescribed Burns, 2000–2007 87
6.1. Diet of Rattlesnakes and Copperheads in Park 129
6.2. Bird-banding data from Tremont and Purchase Knob 141
6.3. Movements of relocated black bear #75, July 1988–July 1990 158
7.1. Federally Endangered and Threatened Plant and Animal Species
     in Park 162
9.1. New Species and New Genus of Invertebrates Named in Honor of
     Arthur Stupka 188
10.1. New Species Recorded for Park, 2000–2007 199
11.1. Predator Beetle Releases in the Great Smoky Mountains National Park . . . 229

Foreword

The Great Smoky Mountains have long been regarded as one of our country's greatest natural resources. The scenic grandeur of these mountains has attracted countless visitors from throughout the world to experience the extremely diverse vegetation communities and abundant wildlife populations. Establishment of the Great Smoky Mountains National Park in 1934 represented an essential step in preserving the wilderness status of this important segment of the Great Smoky Mountains for future generations. The dedicated efforts of the park's staff since its inception have achieved this objective, while greatly enhancing our understanding of the park through a very active program of research and educational outreach. We have come to fully appreciate the special nature of the park, which has been designated an International Biosphere Reserve and a UNESCO World Heritage Site in recognition of its rich biological diversity, particularly within the majestic stands of old-growth forests.

Dr. Donald Linzey has succeeded admirably in developing an especially comprehensive and fascinating account of the natural history of the park. His journey in preparation for undertaking this project has spanned five decades. He initially served as a seasonal park ranger–naturalist and as a graduate student pursuing his Ph.D. research during the 1960s. Linzey's long and distinguished career as an educator, researcher, team leader, and author has maintained a strong focus on the park's wildlife populations. All of these endeavors have instilled in Dr. Linzey a deep understanding of the biology of the park and a strong commitment to assist in the preservation of its natural resources. This book, his tenth, succeeds in this regard through its excellent educational value for scientists, officials, and especially the general public. He employs an informative and readable style of presentation throughout the book, which any reader will appreciate.

The book begins with a historical account of the geological formation of the Great Smoky Mountains and the climatic variations that have resulted at different elevations. Linzey discusses the progression of Native American and European settlements that have occurred within these mountains, including an informative consideration of the individuals and organizations involved in the formation of the park. He adds a nice personal touch to these and subsequent chapters by sharing his most memorable personal experiences within the park over many years. The two major chapters on plant and animal species prove highly educational as well as entertaining. They include the very best of the ecological stories conveyed by park naturalists during organized hikes along many trails. The final six chapters set the book apart from others of its type by focusing on environmental problems facing the park and the important research efforts under way to limit their consequences. The reader has the opportunity to learn about the conservation efforts involved in reintroducing and protecting endangered species populations. The author enlightens us about the earlier development of a long-term research program. This research has now culminated in the ongoing All Taxa Biodiversity Inventory (ATBI), which is attracting national attention for its recognized importance as a model approach. Linzey concludes this section with a rather surprising and somewhat chilling account of current air pollution levels within the park and the many exotic pest species that are seriously impacting major tree species populations. The potential magnitude of these problems is also considered within the context of global warming. Linzey's excellent job of presenting this material on environmental problems should be read carefully by anyone possessing a serious interest in the park.

Don Linzey's book is the most comprehensive treatment available on the Great Smoky Mountains National Park. It is written from the perspective of a highly skilled scientist and naturalist who knows the park extremely well. This combination makes it special as a source of reference on almost any aspect of the park. Most of us grab a few field guides in our specific field of interest and head into the park for another enriching experience. My strong recommendation is to keep your copy of this book close at hand before, during, and after each trip. Doing so will provide the answers to almost any question that arises, point you in new directions of inquiry about species and habitats, and enable you to develop a much broader understanding of what makes the Great Smoky Mountains such a unique and special place. The enlightenment afforded will be worth the effort!

Donald J. Shure, Ph.D.
Professor of Biology, Emeritus
Emory University

Acknowledgments

This book could not have been completed without the assistance of many individuals. Foremost among them is Annette Hartigan, librarian at Great Smoky Mountains National Park, who provided both published and unpublished sources of information from the library and from the park archives. She was always there to answer questions, contact the appropriate person who could provide an answer, and provide advice. She also provided the historical photographs.

The following individuals graciously agreed to review portions of the manuscript in their particular field of interest: Harry Moore (geology), Kristine Johnson and Will Blozan (forestry), Patrick Rakes and Dave Etnier (fishes), Ben Cash (amphibians and reptiles), Ted Simons and Paul Super (birds), Kim DeLozier and Bill Stiver (wildlife), Becky Nichols (ATBI), Janet Rock (plants), Jim Renfro (air pollution), Jeanie Hilten (DLIA), and Keith Langdon (science in the park). In addition, the following individuals provided a variety of information: Bob Wightman (NPS; park acreage); Adriean Mayor (NPS; natural history collection); Jeremy Lloyd, Charlie Muise, and Amber Parker (bird banding data and Tremont historical data); Steve Kemp (GSMA; tree data); Lori Sheeler (East Tennessee Health Department regional epidemiologist) and John New and Nancy Zagaya (University of Tennessee College of Veterinary Medicine; bat rabies); Will Blozan (Eastern National Tree Society; tree data) and Michelle Prysby (Virginia Cooperative Extension) and Jason Love (Tremont; monarch butterfly data). Sharon Williams and Janice Pelton, NPS secretaries in Resource Management and Science, were most helpful in providing annual reports and a variety of other information. Art Stupka's daughters, Carolyn Murrell and Maryann Stupka, reviewed the section on their father.

Many of the photographs used in this book were taken by Steve Bohleber, a DLIA board member from Evansville, Indiana, who spent many hours seeking out and photographing specific plants and animals at my request. Maryann Stupka provided photographs of her father. Mrs. Shirley De Foe and her son, Jay, provided the photo of Don De Foe. Tim Cruze (NPS) provided access to the park's slide collection. Others providing photographs include Jeremy Lloyd, Will Blozan, Ken Jenkins, Ann and Rob Simpson, and Rex Lowe. The bedrock geology, vegetation, and disturbance history maps of the park were provided by Michael Kunze and Ben

Zank. Drawings were prepared by Laurie Taylor. At Wytheville Community College, Anna Ray Roberts assisted with interlibrary loans, while Jerri Montgomery and Shivaji Samanta were always available to help with any computer problems. Jerri Montgomery also assisted in preparing the index.

My wife, Nita, read the manuscript during its development and offered many helpful suggestions. Finally, my sincere thanks to Dr. Don Shure, Professor Emeritus of Biology at Emory University, for critically reading the completed manuscript and offering constructive comments for its improvement. Dr. Shure's extensive ecological research in the southern Appalachians for more than 25 years makes him well qualified to review a work such as this.

Park sign at Gatlinburg entrance.

Introduction

I shall continue to go to the forests and walk in them and listen and fill myself with their greatness and quiet.

—Roger Caras (1979)

My association with Great Smoky Mountains National Park goes back more than 60 years. My parents visited the park in September 1940, leaving their one-year-old son (me) with his grandparents. They were in the park on September 17—just two weeks after the park had been dedicated by President Franklin D. Roosevelt. Their trip was chronicled and illustrated in a scrapbook prepared by my mother. Ever since I can remember, I developed a fascination for the park and a burning desire to see it.

My first visit, in June 1957, was my high school graduation gift from my parents. Although my parents, sister, and I spent only two days in the park, this trip served as the foundation for a lifetime of research, writing, and teaching in the park which now spans 50 years and counting.

In August 1962 my parents brought my sister and me to the park for a second visit. This visit again lasted just two days, but it was a keynote event in my life because I was able to meet Arthur Stupka for the first time (see chapter 9).

I reported for duty as a seasonal ranger–naturalist in the park on June 5, 1963 (see chapter 9) and was employed by the National Park Service for the next two summers. During my Park Service employment, I undertook the mammal research that resulted in my Ph.D. thesis in 1966 from Cornell University. I have continued to engage in mammal research in the park from that time until the present. During the ensuing 43 years, I have obtained several research grants, authored or coauthored two books and 10 scientific papers on park mammals in regional and national journals, produced a number of unpublished research reports, given numerous illustrated programs on various aspects of mammalian biology and natural history, been the subject of a program on *The Heartland Series*

(WBIR-TV, Knoxville) documenting my cougar research, and been the subject of several newspaper articles concerning my research activities.

In 1978 I coauthored *Mammals of Great Smoky Mountains National Park*, which was revised and updated in 1995. The 1978 book was dedicated to Arthur Stupka. Both of these popular books were accompanied by lengthy scientific articles in the *Journal of the Elisha Mitchell Scientific Society* (now the *Journal of the North Carolina Academy of Science*), giving all known data—such as measurements, weights, reproductive data, parasites, and so on—in more detail than the casual reader would have been interested in reading.

In May 1984 my wife, Nita, and I spent our honeymoon in the park. We accessed the Blue Ridge Parkway at Fancy Gap in Virginia (Mile 200) and drove to Cherokee (Mile 469) on our way to the park, where we spent the next four days hiking, observing wildlife, and taking in the beautiful scenery.

In 1998 I accepted the chairmanship of the Mammal Taxonomic Working Group (TWIG) of the All Taxa Biodiversity Inventory (ATBI) (see chapter 10). Our efforts have been devoted to refining the ranges of mammal species in the park, while also seeking possible new species. As chairman, I have produced Web-page accounts for all 70 mammal species currently inhabiting the park or which inhabited the park during historic times. These accounts can be accessed at discoverlifeinamerica.org/atbi/species/animals/vertebrates/mammals.

Since 2003 I have been bringing groups of students and educators to the park to participate in a Natural History Consortium held at the Great Smoky Mountains Institute at Tremont near Townsend, Tennessee (see chapter 12). We join with groups from several other colleges and universities for a week of ecological and environmental sessions that serve as an introduction to the history and problems of the park.

In June 2005 my wife and I purchased a home on the edge of Gatlinburg. Our home is bordered on two sides by the park, so that we now have a "backyard" of 521,257 acres!

Thus, my continuing relationship with Great Smoky Mountains National Park now spans 50 years. During this time, I have not only contributed significantly to our knowledge of the mammalian fauna, but I have also accumulated a great deal of knowledge about the biodiversity of this unique region. This book is designed to give readers an insight to the flora and fauna of the park, together with anecdotes and experiences of a longtime naturalist.

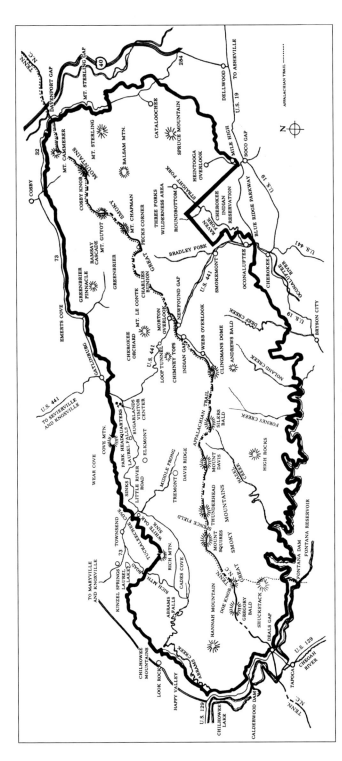

Great Smoky Mountains National Park. From *Birth of a National Park in the Great Smoky Mountains* by Carlos C. Campbell (Knoxville: Univ. of Tennessee Press, 1960).

CHAPTER 1

THE BEGINNING

Mountains are the beginning and the end of all natural scenery.

—John Ruskin (1856)

In the beginning there was only sky and water, sky above and water below. And in all that great wide sea there was no earth, not even a tiny speck of land.

All of the animals lived above the sky, but they needed more room. They looked down upon the water and wondered what was below it, but none of them knew. At last the little water beetle agreed to go down and find out.

He skated over the water, looking everywhere for a solid place to rest. There was none. So he dove deep down into the wide, wide sea and came up with a tiny bit of mud which grew until it became the earth. Later the earth was fastened to the sky with four great ropes to hold it in place.

The newly made earth was very flat and soft and wet. From time to time the animals sent birds out to see if the earth was dry yet. But each time they came back saying the earth was still too wet.

Finally, they sent the great buzzard down to have another look. He flew all over the earth, searching without success for a dry place to land. As the great buzzard became tired, he flew lower and lower until his huge wings began to strike the ground.

Wherever they struck the soft earth a valley was made, and where they raised again a mountain was formed. When the animals saw what was happening they called him back to

Do You Know:
The estimated age of the earth?
Which are the earliest fossilized living organisms?
What the term "plate tectonics" means?
When and how the Appalachian Mountains were formed?
What is meant by "karst"?

the place above the sky. They were afraid that the whole world would be nothing but mountains.

This is the legend by which the older Cherokee Indians explained how the world was created. It also explained how their own homeland was made, for the mountains raised by the beating wings of the buzzard are none other than the Great Smokies.

Geologists tell a different story about the creation of the Earth and the formation of the southern Appalachians. They say that at least three mountain ranges have occupied the region where the Great Smokies stand today. The African continent crashed into the coast of eastern North America about 270 million years ago and crumpled the earth into mountains and valleys. The two continents later separated. Even earlier, other land masses had impacted the eastern edge of the North American continent and left similar mountains. The last impact 270 million years ago left the mountain configuration generally as it is today, except that the forces of weathering such as freezing, thawing, and erosion have rounded their peaks and lowered their elevations.

No one witnessed the Great Buzzard or the continental collision, but geologists can present physical evidence to support their position. Geologists estimate that the Earth formed some 4.6 billion years ago. The oldest rocks discovered on Earth are dated at 3.85 billion years, with the earliest fossilized living organisms being marine microbes (photosynthetic bacteria) that were found in rock from western Australia dated at 3.3 to 3.5 billion years ago. Determining the age of rocks and fossils is done by radiometric dating. A radioactive material undergoes decay, or loss of mass, at a regular rate that is unaffected by most external influences such as heat and pressure. When new rock is formed, traces of radioactive materials are captured within the new rock and held along with the decay product into which it is transformed. By measuring the ratio of decay product to remaining isotope, paleontologists can date the rock and thus date the fossils they contain.

Since its inception, Earth has been undergoing continuous geological changes. Some of these processes, such as volcanoes and earthquakes, are evident and easily observed. Volcanoes had sometimes been active in the region where the Smoky Mountains now rise. They had poured molten material into cracks in the ancient sedimentary stone, leaving intrusions of quartz that today may be seen as narrow white bands in boulders that have tumbled down from the mountains. In addition, the Earth's crust, which is more flexible than we might imagine, consists of seven major, rigid, slablike plates up to about 60 miles (100 km) thick that float on the underlying mantle. These plates (Pacific, African, Eurasian, Australian, North American, Antarctic, and South American) are constantly in motion due to a process known as seafloor spreading, in which material from the mantle arises along oceanic plates and pushes the plates apart, forming rift zones such as the

Midatlantic Ridge. Where they converge, one plate may plunge beneath another, forming seduction zones. Plates also may move laterally past one another along a fault. Because the plates may move up to six inches (15 cm) per year, their movement must be measured with sophisticated devices such as lasers. Radio telescope arrays help to provide data about the movement of the plates that is accurate to within a fraction of an inch. The arrangement of these plates and their movements is known as plate tectonics. The movement of these plates and the continents has significantly affected climates, sea levels, mountain building, and the geographic distribution of life forms throughout time.

To understand the story of the earth, geologists have divided its history into various units of time. The greatest units are called eras. Eras in turn are divided into periods, and the later periods into smaller units known as epochs (Table 1.1). Standard practice is to date the eras, periods, and epochs as millions of years prior to the present time.

The first great era, the Archean, covers the time when life is believed to have begun. It began with the formation of the Earth 4.6 billion years ago and lasted for 2.1 billion years. It covers some 45 percent of all geologic time. The Proterozoic Era, which covers 43 percent of all geologic time, extended from 2.5 billion years ago to 543 million years ago. By the end of the Proterozoic Era, multicellular animals had evolved, as evidenced by fossilized burrows and skeletonized remains. In addition, impressions from soft-bodied forms similar to jellyfish (Ediacarans) are present in late Precambrian fossils (580–543 million years ago) from South Australia, Canada, and elsewhere. Otherwise, we know very little about life from these two eras.

Our knowledge of life on Earth gathered from fossil evidence is essentially confined to the last three eras: the Paleozoic, or time of ancient life; the Mesozoic, or era of middle life (better known as the Age of Reptiles); and the Cenozoic, or era of recent life (the Age of Mammals).

The Paleozoic Era spanned the period from 543 million years ago to 252 million years ago and covers some 6.5 percent of geologic time. It is subdivided into the Cambrian, Ordovician, Silurian, Devonian, Carboniferous, and Permian periods. During the Paleozoic Era, marine sediments continued to be deposited. North and west of the current park area, sediments were being deposited on a slowly submerging continental shelf. A narrow inland sea, created by the sinking of the land, formed west of the present mountains and extended from Canada to Alabama. As the mountains were gradually worn down and became less steep, the flow of the streams decreased in rapidity, transporting finer and finer materials, and depositing them, layer upon layer, at the foot of the mountains and in the sea, the most recent upon the top. Gradually, the inland sea, no longer muddied by the stream-borne sediments, became clear and suitable for primitive marine life of many kinds—trilobites, shelled brachiopods, snails, sponges, and worms. These living organisms thrived and died in the ancient sea, leaving their calcareous shells to settle to the bottom, where they accumulated in vast numbers, forming thick beds of lime. Slowly these beds hardened into limestones, resting upon the older strata of sandstones and conglomerates that had been laid down during pre-Cambrian time. Thus, while most of the Cambrian rocks were composed of shale, siltstone, and sandstone, beginning late in the Cambrian period and extending through the mid-Carboniferous period, carbonate rocks (limestones) became increasingly abundant. By the end of the Paleozoic Era, thousands of feet of sediments had been deposited which now form the Ridge and Valley Province, which extends from New York and Pennsylvania to northern Georgia and Alabama. Only about 10 percent of the rocks in the park were formed during the Paleozoic Era, and they are found only in a few areas such as Cades Cove, Chilhowie Mountain, and Green Mountain.

The Mesozoic Era covers some 3.9 percent of geologic time, beginning some 252 million years ago and ending some 65 million years ago. It is subdivided into the Triassic, Jurassic, and Cretaceous periods. The Cenozoic Era, which began some 65 million years ago and encompasses the present, makes up 1.4 percent of geologic time. During the Cenozoic, tectonic movements shaped the continents into the dispersed forms we observe today.

From Cambrian through Silurian times (543–418 million years ago), most paleogeologists agree that six ancient continents probably existed. These primitive blocks of land were known as Laurentia (most of modern North America, Greenland, Scotland, and part of northwestern Asia); Baltica (central Europe and Scandinavia); Kazakhstania (central southern Asia); Siberia (northeastern Asia); China (China, Mongolia, and Indochina); and Gondwana (southeastern United States, South America, Africa, Saudi Arabia, Turkey, southern Europe, Iran, Tibet, India, Australia, and Antarctica).

Due to the continuing movement of the plates forming the Earth's crust, the land masses collided to form supercontinents and then split apart, enabling new oceans to form. The continental land mass known as Laurentia collided with Baltica between 418 and 380 million years ago, forming a supercontinent known as Laurasia. Between 360 and 252 million years ago, Laurasia collided with Gondwana, thereby forming the world continent Pangaea. Pangaea, the result of multiple collisions that took place over many millions of years, consisted of a single large land mass extending northward along one face of the Earth from near the South Pole to the Arctic Circle. Pangaea was not static; it slowly drifted northward from Carboniferous through Triassic times, causing climatic changes in various areas.

The continental collision forming Pangaea resulted in major mountain building in North America and Europe, including the formation of the Appalachian Mountains approximately 270 million years ago. Thus, the Appalachians were already ancient when the first dinosaurs appeared in the Triassic (approximately 220 million years ago) and flowering seed-bearing plants (angiosperms) first appeared in early Cretaceous (approximately 140 million years ago).

Table 1.1. Geological time scale.

Duration in Millions of Years	Era	Period	Approximate time since beginning of each interval in millions of years before the present
65	Cenozoic	Quaternary Recent (Holocene) Epoch (0.01) Pleistocene Epoch (1.7)	1.8
		Tertiary Pliocene Epoch (5.4) Miocene Epoch (23.8) Oligocene Epoch (36.7) Eocene Epoch (57.9) Paleocene Epoch (65)	65
180	Mesozoic	Cretaceous	142
		Jurassic	200
		Triassic	252
300	Paleozoic	Permian	290
		Carboniferous Pennsylvanian (323) Mississippian (354)	354
		Devonian	418
		Silurian	443
		Ordovician	490
		Cambrian	543
2,000	Proterozoic		2,500
4,100	Archean		4,600

Source: Adapted from Luhr (2003).

Events of mountain-building with folding, faulting, intrusions, and metamorphism are referred to as orogenies. In eastern North America, three orogenies occurred after Precambrian times. Ancient mountains arose from a collision as early as the Ordovician Period (490–443 million years ago) (the Taconic Orogeny). More violent activity occurred later in Devonian-Mississippian times (418–323 million years ago; the Acadian Orogeny) and in Pennsylvanian-Permian times

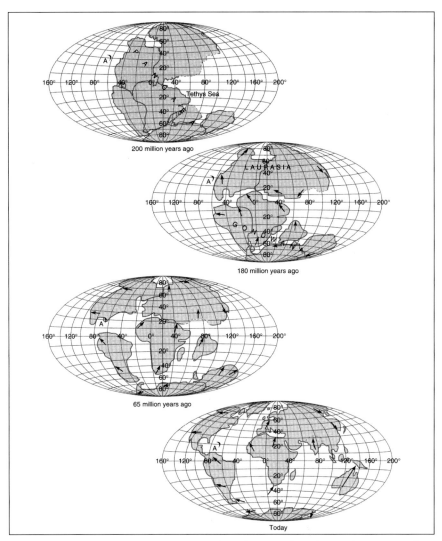

The current position of the continents is not where they have always been. The continents have drifted over the past 200 million years from an original single land mass (Pangaea) to their present positions. Pangaea separated into two supercontinents known as Laurasia and Gondwana, which later broke up into smaller continents. The arrows indicate vector movements of the continents. The black crescent labeled "A" is a modern geographical reference point representing the Antilles arc in the West Indies. From *Vertebrate Biology* by D. W. Linzey (New York: McGraw-Hill, 2001).

(323–252 million years ago; the Alleghenian Orogeny). Together, these Paleozoic orogenies are considered to be three phases of the major Appalachian Orogeny.

During the Permian and Triassic, the eastern part of what is now North America was in contact with Europe and Africa, and South America was joined to

Africa. The higher latitudes were relatively warm and moist during much of this period, while the lower and middle latitudes were probably much drier. Regional differences in rainfall and temperature, as well as the formation of the Appalachian Mountains, led to the development of specific associations of plants (floras) and animals (faunas).

During the Triassic, Pangaea began splitting apart into separate continents, marking the beginning of the independent development of regional biotas. This was followed in the early Jurassic by the beginning of a westward movement of North America away from Africa and South America, although North America still was connected to Europe in the north. This separation, which began in the Jurassic, continues today. Sea-floor spreading is causing the continents to move away from each other at a rate of up to six inches (15 cm) per year. The western movement of the North American continent together with sea-floor spreading in the Pacific Ocean caused rock formations that became the Sierra Nevada, Andes, and Rocky Mountains to be shoved into the continental interior from the west during the Jurassic and Cretaceous periods. Thus, these mountains are much younger than the Appalachians. By the late Cretaceous, North America had moved so far to the west that it was separated fully from western Europe, but it had made contact with northeastern Asia to form the Bering land bridge in the region of Alaska and Siberia. This movement of the North American Plate opened the Atlantic Ocean 80 million years ago.

The continuing separation of North America from Africa and South America allowed the formation of the present-day Atlantic Ocean. As the North American continent moved northward, the climate of the Great Smoky Mountains region changed from tropical to subtropical and then to temperate. During the Mesozoic Era, the recognizable continental outlines of the present world began to form. A massive regression of epicontinental seas (seas covering portions of continents) occurred due to drying conditions in the late Cretaceous, resulting in the exposure of a great deal of land.

The Great Smoky Mountains extend from the Little Pigeon River on the northeast to the Little Tennessee River on the southwest. The Great Smokies are part of the Unaka Mountains. The Unakas are part of the Blue Ridge Province of the Appalachian Mountains, one of the oldest mountain ranges in the world. The Appalachians extend for nearly 2,000 miles from the Gaspé Peninsula in Quebec to central Alabama. The Great Smokies are but a small portion of that range, but they are among the highest and most rugged mountains in the entire Appalachian system.

There are three major types of rocks: sedimentary, igneous, and metamorphic. Sedimentary and metamorphic rocks are common in the park; igneous rocks are rare. Sedimentary rocks are formed when sediments wash into an ocean basin or when remains of sea life are deposited in horizontal layers and later cemented into

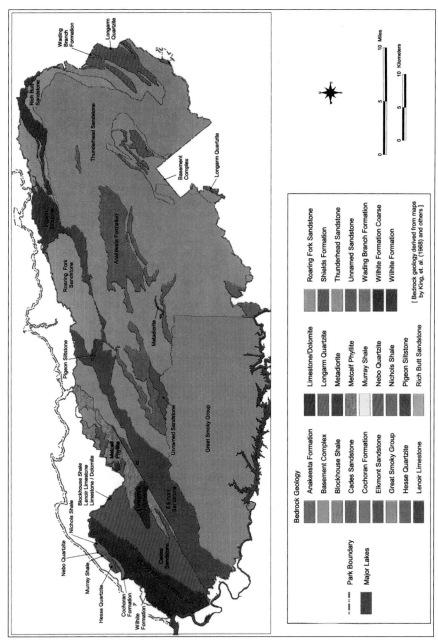

Bedrock geology of the Great Smoky Mountains National Park

stone. Typical examples include sandstone and limestone. Igneous rock, such as granite, forms from molten magma that either cools deep in the earth or is deposited on the surface as lava. Under conditions of heat and pressure, sedimentary and igneous rock can be transformed (metamorphosed) into metamorphic rock. Slate and quartzite are examples of sedimentary rocks that have been metamorphosed.

The infinitely slow but persistent forces of erosion have carved the outline of the Great Smokies as we see them today—rounded summits, jagged peaks, and sheer rock cliffs. Erosion removes the softer shale, slate, and limestone and leaves the more durable sandstone, siltstone, and quartzite behind. Some ridges are formed by ledges of hard rock that resists erosion, with the summits of some being capped by hard quartzite. Some geologists believe that the oldest resistant rock is of Archean age, among the oldest formations known. Lesser ridges are formed of stratified sandstone, conglomerate, and slate that have been broken, tilted, and folded in a variety of patterns. Their once-harsh contours have been worn down by rushing streams, and their rocky skeleton weathered away, forming rich, porous soil where plant life in amazing diversity thrives in an equitable climate. The valleys and slopes are densely clothed with lush vegetation.

At least three mountain ranges have occupied the region where the present Unaka range (Great Smoky Mountains) stands today. The basement rocks that compose the Smokies originated in a sea or ocean that was located between several large drifting continents. About 1 billion years ago these rocks were laid down as sandy, shaly marine sediments in a great trough on the Earth's crust. After the sediment had accumulated to several thousand feet, the trough closed, possibly because of an early collision of the continental plates. The sedimentary rocks were compressed, broken, and folded; molten rock from the Earth's interior invaded and added heat to the process. These rocks, the product of great heat and pressure that changed them from one type of rock to another, were converted (metamorphosed) to a hard crystalline mass consisting of gneisses, schists, and granites that eventually became broken and shoved over one another (faulted). They can be seen in only a few areas in the southern and southeastern portions of the park. One such area is a road cut along U.S. Route 441 between Mingus Mill and Oconaluftee.

Afterward, this area was uplifted, probably due again to plate tectonic forces. The land immediately began to erode, with pebbles, sands, and soil eroding from a very ancient landmass and being deposited as sediment in great quantities on the floor of a shallow inland sea. Over millions of years, these sediments accumulated in layers to a depth of 50,000 feet or more (nine miles) and were gradually changed (metamorphosed) to rock due to their deep burial, compaction, and the chemical action of water. The new deposit solidified between 600 and 800 million years ago into rock that became known as the Ocoee Supergroup. These rocks lie on top of the basement complex. Most of the rock forming the Smoky Mountains consists of the metamorphosed sedimentary rocks that make up the Precambrian Ocoee

Supergroup (Table 1.2). Outcrops of these rocks form Clingmans Dome, Mount LeConte, and the Chimney Tops.

These events were summarized by Moore (1988), who stated:

> As the continents collided, all the rock strata and ocean sediments that were located between the moving continents were crushed, broken, folded, and eventually faulted (broken and shoved over each other). Rock strata which had once been horizontal were tilted at steep angles, folded, fractured, and exposed to extreme pressures and heat generated from the colliding continents. As a result, the rock strata have been metamorphosed (changed to varying degrees from the original material). These metamorphic rocks are classified as metasandstones, metasiltstones, slates, phyllites, and quartzites; all of these are found in the park. The rock strata that were folded and faulted were uplifted, forming a new landmass known as the Appalachian Mountains, the Blue Ridge Mountains, and the Great Smoky Mountains.

The Ocoee Supergroup consists of three groups. From oldest to youngest, they are the Snowbird, Great Smoky, and Walden Creek groups. Although some primitive forms of life had evolved during this period, no fossils have been found in Ocoee rocks. King and Stupka (1950) stated:

> Some of the rocks of the Ocoee series . . . are made up of innumerable pebbles of quartz and feldspar; these pebbles were derived from the breaking apart, under the influence of weather, of individual crystals of an ancient granite mass. The conglomerate looks somewhat like granite and is composed of the same materials but these materials have been broken up, transported, reconstituted in strata, and once more consolidated. The granite from which the conglomerates were derived probably stood as mountain ranges at the time when the Ocoee series was being formed.

Within the park, rocks composing the Snowbird Group are found in the Pigeon River valley on the east side of the park. The major part of the Great Smoky Mountains, however, are formed from the Great Smoky Group, which is made up of three formations—the fine-grained Elkmont Sandstone on the bottom, the coarse-grained Thunderhead Sandstone in the middle, and the dark, silty, and clay-containing Anakeesta formation that has been altered to slate, phyllite, or schist. The hard, resistant Thunderhead Sandstone erodes slowly and forms rounded mountain peaks when it nears the surface. It is responsible for most of the waterfalls in the park. The slaty Anakeesta Formation, however, fractures into thin, jagged pieces and forms the steep-sided ridges and pinnacles of the Smokies. Anakeesta rock is high in sulfide and is often referred to as "acid rock." It is exposed at the Chimneys and at Charlies Bunion along the Appalachian Trail. All of the rock visible along the transmountain road as you travel from Sugarlands to Smokemont belongs to the Great Smoky Group.

As the land surface was raised and subjected to tremendous lateral pressures, it caused the rock formations to buckle into folds and to break in many places. A break in a rock mass along which movement has taken place is known as a fault. A thrust fault occurs when one rock mass is pushed over another rock mass. In most cases, the overriding rock mass is older than the overridden one. The force and extent of thrust faults may shove one rock mass over another over the course of many miles.

As time passed, the strata of pre-Cambrian stone was slowly pushed northwestward, sliding over the top of the more recent limestones. As the tectonic forces pushed the strata

Baskins Creek Falls showing Thunderhead Sandstone formation. Photo by Steve Bohleber.

Chimney Tops in winter. The rocks forming the Chimneys are part of the Anakeesta Formation and are between 600 and 800 million years old. The Cherokees called the Chimneys *Duniskwalguni*, which means "forked antlers."

Table 1.2. Rock types found in Great Smoky Mountains National Park.

Formation	Rock Type
Lower Ordovician Knox Group	Limestone, Dolomite
Lower Cambrian Chilhowie Group	Siltstone, Sandstone, Shale, Quartzite
Ocoee Supergroup (Late Precambrian)	
Walden Creek Group	Shale, Siltstone, Phyllite Coarse conglomerate (containing pebbles)

Great Smoky Group
 	
        Elkmont Sandstone
        Thunderhead Sandstone
        Anakeesta Formation
	Sandstone, Metasiltstone, Slate, Argilite, Phyllite, Schist

Snowbird Group	Phyllite, Siltstone, Sandstone, Quartzite, Arkose
Precambrian Basement	Schist, Gneiss

Source: Adapted from King, Newman, and Hadley (1968).

toward the northwest, they rose, buckling upward in gigantic folds and loops. During this process the older strata (pre-Cambrian rock) were thrust slowly over the newer strata (Basement Complex, Ocoee Supergroup, and Cambrian rocks) as they slid across the land. Thus, in effect, the Great Smoky Mountains have been moved many miles toward the northwest. While most of the rocky skeleton of these great mountains are of pre-Cambrian origin, they now rest upon sedimentary beds laid down at a much later date. As a result, the strata have been reversed, so that older rocks now rest above the newer ones. This momentous event, known as the Great Smoky Overthrust, occurred about 450 million years ago. About 375 million years ago, there followed a period of heat and pressure, and then later more folding and faulting with masses of rock once again being pushed northwest.

Four major faults have been identified within the park: Greenbrier fault, Great Smoky fault, Gatlinburg fault, and the Oconaluftee fault. Most are very high angle faults, except for the Oconaluftee fault, which is a transverse fault. The Greenbrier fault can be seen as one looks eastward from Maloney Point at Fighting Creek Gap. This fault bisects the lower slopes of Mount LeConte and continues on toward the northeast. The Great Smoky fault emerges along the entire northwestern face of Chilhowie Mountain. The Oconaluftee fault stretches along the foothills on the southeastern side of Cades Cove and appears as a break in the slope.

The overthrust of rock formations can be seen in many areas within the park but is especially evident in Cades Cove. In this area, the Ocoee rocks were thrust several miles, overriding much younger formations, mostly limestones, that were formed primarily during the Paleozoic era. The younger rocks contain fossils of primitive marine animals such as trilobites and brachiopods, in contrast to the older rocks of the Ocoee Supergroup which do not contain any fossils. Following the overthrust, relentless stream erosion gradually cut through the ancient rocks and exposed the younger limestones beneath.

All rocks exposed to weather are altered, but their rate of change differs depending on their composition. A region made up of porous limestone containing deep fissures and sinkholes and characterized by underground caves and streams is known as a *karst* area. Since exposed limestone weathers and erodes relatively rapidly, the result is a level-floored valley almost entirely surrounded by steep-sided mountains. Therefore, you can stand in Cades Cove today and be almost encircled by mountains composed of rocks 200 million years older than the rocks of the valley floor. These isolated, flat valleys are usually oval in shape and occur between 1,200 and 1,800 feet in elevation.

This same series of events also formed Tuckaleechee, Miller, and Wear Cove. Karst areas of sinkholes and caves have been created by the chemical weathering of the limestone rock in these areas. Several limestone caves are known to exist in the park, including Gregory Cave in Cades Cove, Bull Cave near Cades Cove,

View of Cades Cove from Gregorys Bald. Photo by Steve Bohleber.

One of four limestone caves in the park, Gregory Cave is located off Cades Cove Loop Road.

and Blowing Cave and Rainbow Cave both in Whiteoak Sink. Such areas provide unique habitats for plants and animals not found elsewhere in the park, and entry is prohibited without a park permit.

Some geologists estimate that when these mountains first formed they may have been 30,000 or more feet above sea level. Currently, the highest point in the Smokies is Clingmans Dome at 6,643 feet above sea level. What happened to the high peaks? Following the Appalachian orogeny, terrestrial weathering forces of rain, sleet, snow, ice, freezing, thawing, erosion, mass wasting (landslides, debris flows), and wind have shaped and formed the topography of the Great Smoky Mountains. Early geologists thought that mountains resulted from the contraction of the Earth's crust as it cooled. According to the plate tectonic theory, however, we now know that mountains are signs not of a shrinking crust but of the growth of the crust. King and Stupka (1950) stated:

> The present ridges and mountains are not caused by upheaval, but by erosion, whereby the valleys have been carved out of the same rock formations as those that still project above them. One may therefore conclude that the landscape of Great Smoky Mountains is not made up so much of ridges rising between the valleys as of valleys cut between the ridges.

The shaping of the Great Smoky Mountains landscape continues today with erosion being the dominant visible force reshaping and redefining the mountain peaks, ravines, ridges, and valleys even while the continents continue to slowly drift. Every year thousands of tons of soil and rock fragments are washed down the slopes and are carried away by the streams. Ice collecting in a crack may chip another piece off the Chimneys. Those rock strata least resistant to weathering—such as shale, slate, phyllite, and limestone—have been eroded away at a faster rate than the more durable rock strata like sandstone, siltstone, and quartzite. The realization that these mountains once stood thousands of feet higher in the geologic past gives one an appreciation of the relentless and powerful effects of erosional forces.

Slide areas such as this one on Mount Le Conte result in the removal of all plants as well as soil.

Soils are the result of erosional forces. They are formed and become differentiated due to a variety of factors, including the climatic temperatures under which they weather, variations in geology, steepness of slope, and the long-term influences of the type of vegetation that grow on them. There is increasing concern about the effects of continuous deposition of acid precipitation on soils at the highest peaks in the southern Appalachians. Biologists know that, generally, as one proceeds up-slope in the park, fewer species are encountered per area, but as you go up, the greater percentage of species are endemics. Endemic species are those with very small geographic distributions, sometimes just a few square miles. The peaks of the Smokies are full of endemic species—plants, lichens, insects, land snails, and even subspecies of vertebrates. Some biologists say that the higher peaks of the southern Appalachians biologically resemble an archipelago of islands. All of these species have at least one thing in common. They all depend on the unique soils that have weathered up there for thousands of years. The impact of acid precipitation on these soils and ultimately on the plant and animal life in the park is discussed in later chapters.

Early settlers avoided the Smokies, largely because the majority of soils, with the exception of some coves and valleys, were poor and rocky. They moved on to the fertile alluvial soils of the Tennessee Valley—where game abounded with the rich dirt. Soils are not only closely connected to the most ancient human settlement and dispersal patterns, but they also remain the foundation of virtually every natural terrestrial system on the planet. Those connections between mineral, soil, vegetation, climate, and animal life are beginning to be better understood in the Great Smokies as a result of an ambitious nine-year soil survey that was completed in 2007. This study, utilizing more than 30 soil scientists from across the United States, resulted in distinguishing 64 soil series throughout the park. Scientists with the U.S. Natural Resources Conservation Service extracted soil profiles known as "monoliths" from a variety of locations throughout the park. In other cases, scientists use "remote sensing" to deduce what likely soil types would be located in a particular area. The study uncovered 20 new types of soils exclusive to the park, most of which are found in the higher elevations. Soils were often given common names reflecting their location, such as Raven Fork, Heintooga, and Peregrine, an organic soil found on the heath balds near Peregrine Peak and Alum Cave Bluff.

Using the soil maps, broken down by area throughout the park, land managers can determine likely vegetation and wildlife type, productivity, chemical content, elevation, percolation ability, and the best sites for potential campgrounds, picnic areas, and roads, as well as the likelihood for notable archaeological resources. Because different soils have differing abilities to neutralize acid, the study will help park managers to pinpoint the streams most vulnerable to acid precipitation. Since soil types directly affect the potential for erosion and landslides, the soil maps will provide critical information for trail maintenance and facility construction. The study has produced digitized soil maps that can be used as overlays in conjunction with park vegetation maps, as well as other resource maps, as part of the park's geographic information system.

CHAPTER 2

TOPOGRAPHY AND CLIMATE

Climb the mountains and get their good tidings. Nature's peace will flow into you as sunshine flows into trees. The winds will blow their own freshness into you, and the storms their energy, while cares will drop like autumn leaves.

—John Muir (1901)

Great Smoky Mountains National Park comprises approximately 800 square miles of mountainous forest. The topography of these mountains is complex, with deep gorges and valleys separated by high ridges. Streams on the northern approaches to the mountains flow northward, while those on the southern approaches flow southward. However, the southward-flowing streams mostly curve westward and then northward just south of the mountains, uniting their waters eventually with those of the northern side and forming the Tennessee River, which drains most of this vast area. As these mountains were thrust upward during the Appalachian orogeny, streams were forced to follow their natural drainage patterns, cutting deep valleys and gorges. The watershed on the south side of Chilhowee Mountain that drains through Happy Valley is the only watershed on the entire perimeter of the park where water flows into the park instead of out. Development in this area raises concerns about runoff and possible septic contamination.

An enormous variety of plant and animal life exists in the Smokies because of the varied topography and range of climatic conditions. These factors produce levels of species diversity unmatched elsewhere in North America. Habitats include five major forest types, meadows, balds, rock

> **Do You Know:**
> How many peaks over 6,000 feet exist in the Smokies?
> The name and elevation of the highest peak in the Smokies?
> How much rainfall occurs annually in the Smokies?

The Little River at Metcalf Bottoms is a popular area for fishermen.

outcroppings, caves, rivers, streams, temporary ponds, and bogs. Approximately 2,115 miles of streams and rivers flow freely within the park.

The Great Smoky Mountains are just a portion of the Appalachian Range, but they consist of some of the highest peaks (Table 2.1). The crest of the Great Smokies runs in an unbroken chain of peaks that rise more than 5,000 feet for over 36 miles. Elevations in the park range from 857 feet (Abrams Creek) to 6,643 feet (Clingmans Dome). There are 16 peaks more than 6,000 feet high, with Clingmans Dome being the highest. It is the highest point in Tennessee, the third highest point in the Appalachian Mountain range, and the third highest point east of the Mississippi River. Only Mount Mitchell (6,684 feet) and Mount Craig (6,647 feet), both located in Mount Mitchell State Park in western North Carolina, are higher. Mount LeConte towers to 6,593 feet from a base of 1,292 feet, making it the tallest (but not the highest) mountain in the eastern United States. It rises to a greater height above its base than any known peak in the East.

One of the most prominent peaks in the Smokies is Mount Guyot, named for the Swiss-born geographer who charted much of this range in the mid-19th century. Arnold Guyot, while teaching at Princeton University, spent the summers of 1856–1860 making barometric measurements in the southern Appalachians to establish elevations. Without benefit of accurate maps or cleared trails, carrying a fragile, cumbersome barometer and enough food for a week, he struggled alone to the top of almost every peak in the Smokies to determine its elevation by calculations based on atmospheric pressure. His measurements have proved amazingly accurate, usually within a few feet of those made by modern surveying techniques.

This range in altitude mimics the latitudinal changes you would experience by traveling north or south in the eastern United States, say from Georgia to Maine. Plants and animals common in the southern United States thrive in the lowlands of the Smokies, while species common in the northern states find suitable habitat at the higher elevations.

With an elevational range of 857 feet to 6,643 feet (a difference of 5,786 feet, a little more than a mile) from the bottom to the top of the Smoky Mountains, temperature and precipitation vary widely (Table 2.2). As a rule of thumb, temperature decreases two to three degrees Fahrenheit for every 1,000-foot increase in elevation and is the equivalent of moving 250 miles northward. Precipitation ranges from an annual average of 55 inches (1.4 m) in Gatlinburg to 85 inches (2.2 m) annually on Clingmans Dome. More rain falls in the Great Smokies than anywhere else in the country except the Pacific Northwest and parts of Alaska. During wet years, over eight feet of rain may fall in the high country. There are many quick showers all year, with periods of precipitation being fairly evenly distributed throughout the year. The park's abundant rainfall and high summertime humidity provide excellent growing conditions. The relative humidity in the park during the growing season is about twice that of the Rocky Mountain region. Autumn (September–October) is usually the driest period. In winter, there is generally a light snowfall at the lower altitudes and a fairly heavy one in the high country.

Seasons in the Smokies vary considerably in the amount of precipitation and in the extent and intensity of heat and cold. Spring is the beginning of the long and varied blooming season, although the months of March, April, and May bring unpredictable weather. Changes can occur rapidly with sunny skies changing to snow flurries in just a few hours. March is the month with the most radical changes, with snow being a possibility at any time during the month, especially in the higher elevations. Warm springlike temperatures, however, may occur as early as January. From mid-April to mid-May, milder temperatures and afternoon showers allow the spring ephemeral wildflowers to bloom profusely in the deciduous forests during a brief window of readily available sunlight and rapid growth before trees leaf out and shade the forest floor. The annual Wildflower Pilgrimage is always scheduled for the last weekend in April in hopes that temperatures have warmed sufficiently

Table 2.1. Top ten peaks in Great Smoky Mountains National Park.

Peak	Elevation in Feet
Clingmans Dome	6,643
Mount Guyot	6,621
Mount LeConte	6,593
Mount Buckley	6,580
Mount Love	6,420
Mount Chapman	6,417
Old Black	6,370
Luftee Knob	6,234
Mount Kephart	6,217
Mount Collins	6,118

Table 2.2. Average monthly temperature and precipitation data for Gatlinburg, Tennessee, and Clingmans Dome. Temperatures are in degrees Fahrenheit.

	Gatlinburg, Tennessee Elevation 1,462 feet			Clingmans Dome Elevation 6,643 feet		
	Avg. High	Low	Precip.	Avg. High	Low	Precip.
January	51	28	4.8"	35	19	7.0"
February	54	29	4.8"	35	18	8.2"
March	61	34	5.3"	39	24	8.2"
April	71	42	4.5"	49	34	6.5"
May	79	50	4.5"	57	43	6.0"
June	86	58	5.2"	63	49	6.9"
July	88	59	5.7"	65	53	8.3"
August	87	60	5.3"	64	52	6.8"
September	83	55	3.0"	60	47	5.1"
October	73	43	3.1"	53	38	5.4"
November	61	33	3.4"	42	28	6.4"
December	52	28	4.5"	37	21	7.3"

Weather Summary. Published by Great Smoky Mountains National Park, December 2000.

to permit the emergence and blooming of the park's many spring wildflowers. The average daily maximum and minimum temperatures during March, April, and May show a rapid and steady rise. Due to the range of elevations, however, colder, winterlike conditions may still exist at high altitudes even into May, whereas almost summerlike conditions may exist in lower elevation areas by this time. On April 16, 2007, LeConte Lodge recorded a low of 1 degree F, an April record.

The months of June, July, and August are the hottest and wettest months of the year. Summer in the Smokies means heat, haze, and humidity. Brief afternoon or evening thundershowers often occur. At the lower elevations, temperatures range from warm to hot during the day, but generally cool during the evening and overnight. Even during these summer months, high-elevation areas are generally cool and require the use of blankets or sleeping bags by hikers or campers. On Mount LeConte (elevation 6,593 feet), no temperature above 80 degrees has ever been recorded. Biting midges and gnats may often be a problem during the summer months, especially near streams and damp places. Among the summer-blooming plants are yellow-fringed orchid, Deptford pink, and fire pink. From mid-June to mid-July, spectacular flower displays of flame azalea, rhododendron, mountain laurel, and other heath family shrubs occur, especially on high-elevation heath balds.

The showy orchis generally blooms during April and May. Although small in size, the velvety purple and white flowers are very attractive.

The inflated yellow pouch of the yellow lady's slipper is unmistakable. The surface of the slipper is waxy-textured and can give the impression of freshly sculpted, fine porcelain. Photo by Steve Bohleber.

Pink lady's slipper. Photo by Steve Bohleber.

Deptford pink is a slender, somewhat rigid plant that generally stands 8 to 12 inches tall and blooms from June to August. The light to dark pink flowers possess small white spots.

September, October, and the first half of November are the driest months. The autumn color season begins with clear skies and cooler weather. This is the time of year when the atmosphere is the clearest. While warm temperatures may be experienced during early autumn, daytime highs may only be in the 50s and 60s by early November with the possibility of snow in the higher elevations. The first frosts often occur in late September, and by November the lows are usually near freezing. Much of the park's visitation occurs during autumn as visitors come to see the colorful displays of fall foliage.

Leaves of deciduous trees contain several different pigments—chlorophyll, carotene, and xanthophylls. Together these pigments make up only about 3 percent of

Autumn colors in Sugarland Valley are vivid when viewed from Big Walnut Overlook.

a leaf's weight, yet they are responsible for all of its color. Chlorophyll is responsible for the green color of leaves during the summer months. Cool nights and bright days during autumn bring about physiological changes in the leaves that result in a change of coloration. Sourwoods, dogwoods, and sumacs turn scarlet and crimson. Poplars and hickories turn golden-yellow. Persimmons are royal purple. Scarlet oaks become a flaming crimson. Carotenes are responsible for yellowish-orange and red, while xanthophylls impart yellow colors. Carotenes and xanthophylls are present in the leaf all summer, but at the approach of autumn a new class of pigments—anthocyanins—appear. Anthocyanins are responsible for brilliant hues of scarlet, purple, or even deep blue. The colors of many flowers are also due to anthocyanins—the scarlet of cardinal flowers, the deep red of roses, the blue color of violets and chicory. The bluish color of grapes and blueberries and the red color of ripe apples are also caused by anthocyanins. Another pigment, tannin, is found in many leaves and is responsible for the brown or golden-bronze colors of some oaks.

Light and temperature are critical factors in producing fall foliage coloration. Light enables a leaf to produce more sugar which, in turn, affects pigment formation. Temperature is of paramount importance, but the most favorable range for color production is slightly above freezing. Freezing temperatures, which trap sugars and tannins in the leaves, simply kill the leaves before color is produced. Thus, cool but not freezing nights are most favorable. As a general rule, most bright autumn colors are especially evident during years that are sunny and dry, followed by rain in early fall and, later, by cool, but not freezing, nights. Cloudy, warm autumns usually result in dull colors, since sugar production is reduced.

The fall months may also be the time when the remnants of hurricanes and tropical depressions may affect the park with high winds and flooding. In addition, frontal systems from the west may occasionally bring similar conditions. Such was the case on October 16–17, 2006, when a storm system with a top measured wind gust of 106 miles per hour was recorded on Cove Mountain. This storm resulted in extensive tree and property damage and resulted in the closure of many roads on the Tennessee side of the park.

The winter months from mid-November through February may bring snow—up to several feet at a time in the higher elevations but much less in the valleys. Mount LeConte received 31 inches of snow during a snowstorm on February 11, 2006. This was on top of 10 inches already present, making a total of 41 inches on the ground. Gatlinburg received 8–10 inches of snow from this same storm. Snowfalls of more than one inch generally occur three to five times per year at the lower elevations. Many winter days have high temperatures of 50 degrees or more. High temperatures occasionally even reach the 70s. On January 1, 2006, for instance, my wife and I were hiking in shirt sleeves in temperatures approaching 60 degrees F. Some hikers were in shorts. We again experienced 60 degrees F on January 27, 2007. Most winter nights have lows at or below freezing. Wider vistas are now possible because of the winter defoliation of the deciduous trees. Frozen waterfalls can be spectacular.

In December 2004 my wife and I began hiking up Low Gap Trail from Cosby Campground. There was about one inch of snow on the ground, and the hoarfrost, winter's equivalent to summertime's dew, had created a frozen winter wonderland by outlining every branch and twig in white. We had only planned on hiking a short distance, but the scenery was so breathtaking that we ended up hiking all the way to Low Gap. I cannot remember a more beautiful winter hike in all my years in the Smokies.

Is there any one best time to visit the Smokies? Each season has something to offer, and everyone has his or her own preference. Many prefer spring for its

Winter is a quiet time along Cades Cove Loop Road.

spectacular displays of wildflowers and shrubs, while others prefer autumn with its spectacular and colorful fall foliage. During the colder winter months, snow and hoar-frost create a wonderfully different landscape. Although certain aspects of the Smokies are always changing with the seasons, other features undergo little change. For example, the swiftly flowing, crystal-clear mountain streams flow regardless of the season, although during winter the rocks and boulders may become covered with snow and/or encrusted by ice. The drifting clouds, the starlit skies, the quiet solitude of the backcountry—all are affected little by the change of seasons. To visit the Smokies during only one season makes it difficult to appreciate all that the Smokies has to offer. To really experience the beauty of the Smokies, one needs to visit during all seasons.

During my 50 years in the Smokies, I have been in the park during every month of the year. I have hiked hundreds of miles of trails (many of them numerous times) as well as covering untold miles off-trail in the course of my mammal research. While I have been to Gregory Bald, Andrews Bald, Chimney Tops, Alum Cave, Charlies Bunion, and many of the other well-known sites in the park, I have also experienced the adventure of hiking off-trail and finding beautiful streams, cliffs, wildflowers, old homesites, and much more that most visitors never get a chance to see.

Biological diversity is the hallmark of Great Smoky Mountains National Park. No other area of equal size in a temperate climate can match the park's amazing diversity of plants, animals, and invertebrates. Over 80 discrete vegetation communities exist within the park. Over 10,000 species of organisms have been documented

Monthly Calendar of Natural Events

January
Female black bears give birth to their cubs.
Male white-tailed deer shed antlers.
Winter birds present include ruffed grouse, belted kingfisher, cedar waxwing, yellow-rumped warbler, purple finch, red-breasted nuthatch, pine siskin, black-capped chickadee, winter wren, American goldfinch, barred owl, and wild turkey.
Great horned owls raising young.
Wood frogs and spotted salamanders breeding in ponds and pools.

February
Red maple trees bloom.
Trailing arbutus may bloom along trail edges.

March
Spicebush blooms.
Wildflowers that may bloom this month include spring-beauty, sharp-lobed hepatica, bloodroot, and several violets.
Arriving migratory birds include solitary vireo, yellow-throated warbler, black and white warbler, and Louisiana waterthrush.
Spring peepers and chorus frogs breeding.

April
By early April, most black bears will have emerged from their winter dens.
Flowering dogwood trees reach their peak bloom about mid-month.
Many woodland wildflowers are in bloom, including foamflower, columbine, fire pink, Dutchman's breeches, trout lily, large-flowered white trillium, yellow trillium, wake robin, crested dwarf iris, showy orchis, white fringed phacelia, early meadowrue, large-flowered bellwort, and wild geranium.
Many warblers and other migratory birds arrive to spend the summer and breed. They include ruby-throated hummingbird, veery, wood thrush, yellow-throated vireo, chestnut-sided warbler, Blackburnian warbler, Canada warbler, and scarlet tanager.
Monarch butterflies begin arriving.

May
Mountain laurel in bloom.
Flame azalea in bloom at lower elevations.
Silverbell trees and tuliptrees (yellow poplars) in bloom.
Umbrella and Fraser magnolia trees in bloom.
Woodland wildflowers in bloom include creeping phlox, wake robin, showy orchis, yellow lady's slipper, yellow star grass, galax, fire pink, and bluet.

June
Catawba rhododendron reaches its peak of bloom.
Rosebay rhododendron reaches its peak of bloom at the lower elevations.
Speckled wood lily, galax, bluets, and Rugel's ragwort bloom.
Flame azalea blooming at higher elevations, especially on grass balds.
Elk calves being born.
Black bears breeding.

July
Sourwood trees bloom.
Rosebay rhododendron reaches its peak of bloom at the middle and high elevations.
Wildflowers in bloom include butterfly-weed, yellow-fringed orchid, cardinal flower, purple-fringed orchid, and fly poison.
Goldfinches begin nesting in late July.

August
Pin cherry fruits are ripe.
Wildflowers in bloom include Joe-Pye-weed, Turk's cap lily, pink turtlehead, heart-leaf aster, ladies' tresses, goldenrod, bee-balm, and touch-me-not.

September
Trees showing early autumn color include sourwood, pin cherry, flowering dogwood, and yellow birch.
Flocks of migrating broad-winged hawks may be seen from Clingmans Dome and Look Rock towers.

October
During the first half of the month, fall colors will reach their peak at the high elevations.
Fall colors will reach their peak at the lower elevations during the second half of the month.
Witch-hazel begins blooming with bright yellow flowers.
Monarch butterflies migrating.

November
Oak trees continue to show good color early in the month.
Watch for possible arrival of evening grosbeaks.
The deciduous leaves of buffalo nut remain bright green.
Many fall asters continue to bloom.
Woodchucks and jumping mice begin hibernating in underground burrows.

December
Mammals in various states of deep winter sleep include black bear, woodchuck, chipmunk, and jumping mice.

Adapted from calendar prepared by Great Smoky Mountains National Park.

in the park; scientists believe an additional 90,000 undocumented species may also be present. The glacial influence on the Smokies' climate (chapter 5), coupled with the range of elevation and the southwest-to-northeast layout of these mountains, accounts for the striking variety of living things. The park is almost 95 percent forested, of which roughly a quarter is old growth. It is one of the largest blocks of deciduous, temperate, old-growth forest in North America. Five forest types within the park support 1,660 species of vascular plants and over 490 species of nonvascular plants. The forests of Great Smoky Mountains National Park are world-renowned for their biological diversity.

CHAPTER 3

Pre-Park History

The most common trait of all primitive peoples is a reverence for the lifegiving earth, and the native American shared this elemental ethic: the land was alive to his loving touch, and he, its son, was brother to all creatures.

—Stewart Udall (1963)

The recorded history of impacts on the flora and fauna of the Smokies begins with the Cherokees, the first known inhabitants. They called themselves *Ani-Yunwiwa*, "the Principal People," Cherokee being the white man's name for them. Their lands covered 40,000 square miles including the Great Smokies and parts of what today are the states of Tennessee, North Carolina, South Carolina, and Georgia. The Indians called the Great Smokies *Shaconage,* meaning "place of blue smoke." They lived by the thousands in villages along streams and rivers, many concentrated along the Little Tennessee River that flows along the southeastern boundary of the present national park. The ancient capital town was Echota, situated near the mouth of the Little Tennessee River. But they ranged far across their lands following ancient paths, including a transmountain route that crossed the Smokies at what is now called Indian Gap, just west of Newfound Gap. Through successive treaties beginning in 1761, the Cherokees lost more and more of their territory to encroaching white settlers until they finally gave up their rights to all their remaining land east of the Mississippi River by the treaty of New Echota with the U.S. government in 1835.

Evidence shows that Native Americans had been present in the park since at least 8,000 years before the present. Archaeological excavations undertaken during 2003 and 2004 near Smokemont in the southeastern

> **Do You Know:**
> The earliest date that archaeologists have verified human occupation in the Smokies?
> The first European to see the Great Smoky Mountains?
> The ancestry of most of the early white settlers in the park?

portion of the park as part of an environmental assessment for the rehabilitation of Newfound Gap Road uncovered evidence of human use from about 8,000 years ago to the present. Prehistoric materials from this site appear to show evidence of human use from 6000 B.C. to A.D. 1350 and archaeological remains confirm historic Cherokee occupation of the area. Bog sediments from the Cherokee Indian Reservation adjoining this portion of the park show an increase in charcoal corresponding to the arrival of the Cherokee between A.D. 1450 and 1600. Excavations for a proposed road widening at Greenbrier in the northwestern portion of the park in 2004 uncovered evidence of a continuum of human occupation dating back to 10,000 years ago. Archaeological excavations at Twin Creeks have identified human occupation during the Middle to Late Archaic (6000 B.C.–1000 B.C), the Middle Woodland (700 B.C.–A.D. 200), and during the historic period. Archaeological evidence, such as the recovery of burned clay daub and a charcoal-rich pit feature, suggests the area was utilized semipermanently, and future work could identify the remains of a prehistoric household. The charcoal recovered from the feature will provide a radiocarbon date. Gregory Cave in the western portion of the park has a long history, and evidence has been recovered of Woodland period deposits (1000 B.C.–A.D. 1000). In addition, cane marks and a turkey motif have been tentatively identified on the cave walls.

The southern Appalachians provide a fascinating case study for the transformation from hunting and gathering to farming. The peoples of the Archaic period beginning about 8,000 years ago had a diverse menu of wild mountain plants and hunted game. Over the next 5,000 years they shifted emphasis to wild seedy annuals that are typically found at lower elevations and encouraged sunflower and chenopodium (or goosefoot) to grow in disturbed areas near their camps. By the time maize, or corn, was traded into the American Southeast (it was originally domesticated in northern Mexico), the woodland peoples of southern Appalachia were preadapted for farming. Corn was planted wherever soil was rich and well-watered, often in riverbottoms. The mound-building cultures farmed corn as a major part of their economy. Mound-builder campsites have been identified in the Greenbrier area of the park, although the nearest big sites are found in Sevierville, Tennessee, and Franklin, North Carolina.

During the sixteenth century, Spanish conquerors Hernando de Soto, Juan Pardo, and Tristan de Luna invaded what is now the southeastern United States. De Soto is believed to have been the first white man to see the Great Smokies. He visited the Cherokee town of Tallassee in 1540 on his way northwest to a place named Chisca, near today's Knoxville, where there were supposedly mines of gold. He found the Cherokees—he called them *Chalaque*—a quiet, agricultural people governed by a loosely knit tribal organization. They lived in sturdy, grass-roofed houses with walls of upright poles interlaced with cane and covered with clay. The Cherokees were an agricultural people, farming the valleys; cutting trees for cabins, dugout canoes, and firewood; and hunting in the forests. They raised corn, potatoes, beans, pumpkins, squash, melons, peas, and tobacco on the fertile, broad river bottoms. Although the Spanish invaders did not reach the Smokies, their trade goods, plants, and diseases did. By the mid-1600s, the Cherokees became increasingly committed to European trade. They began hunting more intensively, which led to the disappearance of bison and elk. By the eighteenth century, the Cherokees had begun trading furs with the English, which further reduced the populations of white-tailed deer, beaver, and otter.

Before the American Revolution, the Cherokee discouraged settlers. After the defeat of their English allies, however, they sought peace with the new U.S. government. The Cherokee adapted well. They built modern houses, and attended school, and by 1820 they had created a written language. Despite the Indians' assimilation, many Americans wanted to move all Indians west of the Mississippi River. The discovery of gold on Cherokee lands in 1815, and Andrew Jackson's rise to the presidency, led to their removal and the tragic Trail of Tears. During the winter of 1838–39, the U.S. Army forced more than 14,000 Cherokees to move to Indian Territory (present-day Oklahoma). They were compelled to make the long journey to Oklahoma on foot during the coldest part of the winter. About one-quarter of them died from the hardships that they suffered.

Not all of the Cherokees were driven out. Several hundred refused to move and hid in the Great Smoky Mountains, avoiding the army and local authorities. In the years that followed, this little band was joined by others who walked all the way back from Oklahoma. In the 1870s, the U.S. government allowed these renegade Cherokees, now called the Eastern Band of Cherokees, to claim some of their lands in western North Carolina. Today their descendants live on the Qualla Indian Reservation just outside of Great Smoky Mountains National Park.

The Cherokees were well educated. This was made possible by the invention of an alphabet containing 85 characters. It was invented by Sequoyah, the son of a Cherokee mother and a white father. It took him 12 years to develop his simple but famous alphabet. Thousands of Cherokees learned to read, write, and spell in a very short time. In 1828 a weekly paper, the *Cherokee Phoenix*, was printed in both Cherokee and English. Still later, books were translated into the Cherokee language and printed for use by these people.

The first European Americans to settle within the current boundaries of the park were John J. Mingus and Felix Walker. Their homes were established in the 1790s just north of the present-day town of Cherokee. The date of arrival of the first settlers in Cades Cove is unclear, according to Randolph Shields, a longtime resident and author. John and Lurany Frazier Oliver are thought to have moved into the cove in 1818, although no one could legally own land there until 1819 when the Cherokees relinquished their claim through the Calhoun Treaty. Before their arrival, Cades Cove was part of the Cherokee Nation. The Cherokee called the cove *Tsiyahi* (place of the river otter). In addition to river otters, elk and bison also lived in the cove. Bison were probably extirpated in the late 1700s, while the last elk in East Tennessee was reportedly shot in 1849. The Cherokee never lived in the cove, but they used it as a summer hunting ground.

The first recorded legal title for land in Cades Cove following the Calhoun Treaty was in March 1821 to William Tipton, who was granted the rights to 1,280 acres. In Appendix A of his book on Cades Cove, Randolph Shields lists the 36 land grants that were recorded from 1821 through 1890. Other early settlers included the Cables, Shieldses, Sparkses, Gregorys, Burchfields, Whiteheads, Powells, Jobes, Lawsons, and Ledbetters. Shields noted that only two families—the Shieldses and Olivers—lived in the cove throughout the life of the cove until the land was purchased for the park. By 1840 the population of Cades Cove consisted of 70 families with 451 people. By 1850 the population reached its maximum of 685 people in 132 households. Shields stated that the Methodist congregation may have organized as early as 1824, with the Baptists organizing in 1827.

Although the Cataloochee area had been used by settlers for hunting and livestock grazing, it was not until 1839 that the first permanent homesites were established. These earliest settlers were Evan Hannah, William Noland, and James and Levi Caldwell. By the late 1830s and early 1840s, European American farmers had settled nearly all of the major stream valleys, and for the rest of the 19th century both Cherokees and whites shaped the Smokies as they farmed the lower elevations. Dikes were built to decrease flooding, and wetlands, crucial to many species of plants and wildlife, were drained in order to increase farmland. Farm ponds were created. The settlers moved up the coves and into the river valleys of the mountains. They cut trees to build log cabins and open fields where the farmers then planted corn, the primary crop. They were a self-sufficient people who gathered nuts and berries and hunted wild game while cultivating crops and raising hogs and chickens. The biggest difference in land use brought about by the white settlers was the introduction of domesticated livestock. Some of the livestock, especially pigs, were free-roaming. Cattle and sheep inhabited the valleys during the winter months and were herded to the high-elevation grassy balds during the summer.

Both the Cherokees and the white settlers impacted the ecosystem of the southern Appalachians by using fire. The Cherokees used controlled fires to remove trees and clear vegetation to create farmland and improve hunting. Fire allowed propagation of fire-adapted species such as the pines and good nut- and acorn-bearing trees, exposed soils to germinating seeds, added organic matter to the soil, and killed plant pathogens and harmful insects. White settlers also employed fire to clear understory plants, promote grass and woody sprouts for livestock, kill insect pests, and aid berry reproduction.

Most of the early white settlers were of Scotch-Irish ancestry. "Scotch-Irish" was a term applied to Scots who relocated to Ireland in the early 1600s and whose descendants then began a migration to America in the early 1700s. They first settled the Pennsylvania region, then moved south through Virginia into the Carolinas and northern Tennessee, where they joined with Germans and English to penetrate the mountains. Most of the pioneers who moved into the Great Smokies during Revolutionary War times or soon after migrated from the Watauga Settlement in northeastern Tennessee. As they established themselves, mountain people hunted bear most often to protect their stock. Other large predators such as wolves and panthers were also threatening. Bounties were established. The Codes of Tennessee for 1858 and 1884 stated that hunters would receive $6 for each wolf pelt (over four months old) and $1 for each red fox pelt. A "wild cat" hide could be "received by the tax collector in payment of the poll tax" for one year. By the 1870s, the number of predators had been reduced to such an extent that bounties fell to $2 for a wolf, $1 for a wildcat, and 50 cents for a fox. For the most part, bounties disappeared by 1900 because these animals no longer existed in numbers great enough to threaten cattle or sheep. The wolf bounty, however, remained until 1917.

The earliest mention of the status of wolves in this area came in 1844, when a letter to a member of the House of Representatives stated that sheep were destroyed by wolves, "which have not yet been entirely exterminated." In 1859 wolves were reported to be "troublesome" to the mountain farmers of North Carolina and Tennessee. In letters written to Arthur Stupka, the park's first naturalist, in 1952 and 1953, Dr. G. S. Tennent of Asheville related several instances of wolves in this area. In 1890 a wolf was killed in Cataloochee Township and another killed near Asheville, North Carolina. Tennent also related that prior to the 1890s wolves were plentiful in the Cataloochee Mountains and in the wildest parts of the Balsams, but, with the coming of the railroad, they disappeared. Dr. Tennent concluded, "These facts would put the final disappearance of 'wolves' about the middle eighties, and leave the possibility of one or two strays hanging on into the present century." In 1887 C. H. Merriam noted that wolves "still occur" in the Great Smokies, and John Oliver, a former resident of the park, remembered hearing wolves howling in Cades Cove when he was a boy (1880–1890). D. Ogle of Gatlinburg recalled seeing one of these animals that had been caught in a bear trap near the Sugarlands during the 1890s. He also heard two wolves howling near the area that is now Chimneys picnic area. In 1944 C. S. Brimley wrote that wolves were "apparently finally exterminated in or about 1890, up to which time they still occurred sparingly in the mountains." Finally, W. L. Hamnett and D. C. Thornton stated in 1953: "In the Mountain Region . . . wolves existed in the more remote sections until the 1800's and possibly until the very early 1900's."

As predators were reduced and fields continued to be cleared, many species of fishes, amphibians, reptiles, birds, and mammals were affected. No longer was there continual forest; it was now fragmented into smaller blocks interspersed with fields and human habitations. Without shading, stream temperatures increased. Streams were also polluted by livestock waste and siltation from erosion. Populations of birds such as crows and pigeons increased, as did the numbers of mammals such as opossums, squirrels, and raccoons. The mountaineers ate the meat of deer, bears, squirrels, rabbits, wild turkeys, and ruffed grouse. They sold pelts of mink, opossum, raccoon, muskrat, and fox. Horace Kephart (1913) noted that

> the deer are all but exterminated in most districts, turkeys and even squirrels are rather scarce, and good trout fishing is limited to stocked waters or streams flowing through virgin forest. The only game animal that still holds his own is the black bear, and he endures in few places other than the roughest districts, such as that southwest of the Sugarland Mountains, where laurel and cliffs daunt all but the hardiest of men.

Andre Michaux, the famous French botanist, wrote glowingly of the wealth of azaleas and rhododendrons found in the region in 1793 and again in 1802. William Bartram of Philadelphia came into the Cherokee country, which adjoins the park, about 1776. In 1842 the botanist Dr. Ferdinand Rugel made an excursion into the Smokies. On Mount Mingus he discovered the Smoky Mountain ragwort, named *Senecio rugelia* by Asa Gray. (The name was later changed to Rugel's ragwort or Rugel's Indian plantain [*Rugelia nudicallis*].) It is one of the few flowering plants found only in the Great Smokies. S. B. Buckley, a North Carolina geologist and botanist, also began studies in the Smokies as early as 1842. Mount Buckley, just west of Clingmans Dome, bears his name. Thornborough (1942) related the following:

> In the spring of 1928, the late Dr. H. S. Pepoon, nationally-known botanist, came to the Smokies in search of plant life and surprised even himself by identifying no less than 513 specimens in five days. Gray of hair, but young in enthusiasm, Dr. Pepoon told me on leaving, "Usually one finds about all the specimens there are the first two days, then only an occasional new one, but I averaged almost a hundred a day, and judging from what I found here this week, I should say there ought to be close onto 2,000 specimens."

Dr. H. M. Jennison, a professor of biology at the University of Tennessee, who served as a wildlife technician in the park from 1935 to 1937, listed 1,500 species. (As of January 2008 a total of 1,660 species of vascular plants have been identified in the park.)

Vertebrate studies also began in the early 1900s. Emmet Reed Dunn collected salamanders extensively throughout the mountains of western North Carolina, including the area around Mount Sterling, in 1919. Here he recorded the first North Carolina records for the red-cheeked salamander (*Plethodon jordani*) and for the three-lined salamander (*Eurycea guttolineata*).

The first extensive survey of mammals in the park was undertaken between 1931 and 1933 by E. V. and Roy Komarek from the Chicago Academy of Sciences. Their findings have served as a baseline for subsequent mammal investigations.

One species that they recorded, the marsh rice rat (*Oryzomys palustris*), has never again been found in the park.

A similar baseline study of amphibians and reptiles was compiled by Willis King, an associate wildlife technician, in 1939. In 1936 he published a paper describing as a new species the pigmy salamander that he had discovered on Mount LeConte. Upon his resignation in November 1940, to accept a position with the North Carolina Division of Game and Inland Fisheries, Arthur Stupka wrote, "Our excellent collection of Great Smoky Mountains National Park reptiles and amphibians is largely the result of his activities."

Other significant early scientific research in the park as compiled by Henry Lix (1958) included:

1930	W. H. Weller published an article reporting the presence of the green salamander (*Aneides aeneus*) in the park. He also published the park's first preliminary checklist of salamanders in 1931.
1931	G. W. McClure published a paper on six kinds of salamanders in and around LeConte Creek.
1934	Dr. N. S. Davis of the Chicago Academy of Sciences submitted an application to establish a biological field station in the park. W. L. Necker reported a total of 21 species of reptiles and amphibians.
1935	Dr. H. W. Camp of the New York Botanical Gardens came to the park to study the Ericaceae. Dr. W. M. Barrows collected more than 100 species of spiders during June.

It was inevitable that the great forests with their mammoth trees would attract the attention of the outside world. In the late 1880s, lumber companies began moving into the mountains, bought land, and began cutting roads and laying rail lines to haul out the trees. Captain McDonald managed what was possibly the first logging operation in the park in the 1880s. He cut timber in the Laurel Creek area up to Schoolhouse Gap. Company towns such as Smokemont, Tremont, Elkmont, and Proctor were built to house the workers. The Little River Lumber Company on the Tennessee side and the Champion Fibre Company on the North Carolina side were the largest. The J. J. English Company logged the watershed of the Middle Prong of Little River between 1880 and 1900 and floated the logs down the river to sawmills located at Lenoir City. They selectively cut mostly big tulip trees and left most of the forest standing with little damage. Other lumber companies in the early 1900s were mostly clear-cutters; they cut every tree big enough to saw.

The roads, mills, railroads, and towns destroyed the habitat for many vertebrates. In addition, clear-cutting led to the mountains being devastated by fire and erosion. When trees that once shaded streams were removed, some of the waters became too warm for brook trout. Many of the great chestnuts and oaks that had provided food for bears, wild turkeys, deer, squirrels, deermice, chipmunks, and other creatures were gone. The great quantities of slash and woody debris were left in the mountains where they were dried by the sun and served as fuel for fires that blackened huge areas. The denuded mountainsides were subjected to flooding and extensive erosion which deposited great quantities of silt in streams and rivers at lower elevations. Horace Kephart, an early park supporter, described the following:

> When I first came to the Smokies the whole region was one superb forest primeval. I lived for several years in the heart of it. My sylvan studio spread over mountain after mountain, seemingly without end, and it was always clean and fragrant, always vital, growing new shapes of beauty from day to day. The vast trees met overhead like cathedral roofs. . . . Not long ago I went to that same place again. It was wrecked, ruined, desecrated, turned into a thousand rubbish heaps, utterly vile and mean."

Jim Shelton's family in front of one of the park's magnificent American chestnut trees at Tremont in 1927.

In supporting the creation of a national park, he further stated:

> Why should this last stand of splendid, irreplaceable trees be sacrificed to the greedy maw of the sawmill? Why should future generations be robbed of all chance to see with their own eyes what a real forest, a real wildwood, a real unimproved work of God is like.

Clearcutting near Elkmont resulted in loss of habitat as well as erosion. The lack of tree cover along streams resulted in higher water temperatures, which affected species such as the brook trout.

Elkmont was once a thriving town with houses and cabins, a hotel, machine shop, commissary, coaling dock, and a church.

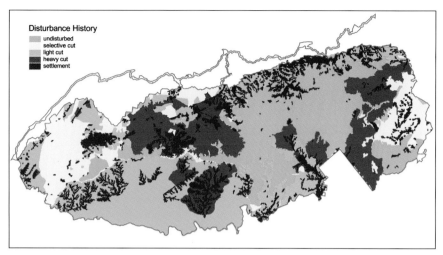

Disturbance history in the Great Smoky Mountains National Park.

Fortunately, Great Smoky Mountains National Park was completed in time to stop the axes before they reached some 200,000 acres or more of forest. These virgin forests are now preserved forever in their natural condition as a precious irreplaceable American heritage.

By the time the Great Smokies passed from private ownership to the National Park Service, the forest, the land, and the fauna were in trouble. Overhunting and illegal hunting, along with the wreckage by logging of more than half their habitat, had left some animals almost extirpated in the park. White-tailed deer were one example. From 1933 to 1941, J. F. Manley, who became the park forester in 1933, reportedly saw only one deer and four bears in his travels throughout the park. Dr. A. Randolph Shields, who grew up in Cades Cove and became a biology professor at Maryville College, reported that only about 30 deer lived in the mountains above the cove during this time,

But with hunting banned by the National Park Service and the forest given a chance to renew itself, wildlife rebounded. Within the next 30 years, 100 or more deer could be counted as one drove around the 11-mile Cades Cove Loop road at dawn or dusk. The bear population also increased, as did the populations of many other wildlife species. Over the years, however, as ecological succession proceeded, populations of some species would decline. Suitable habitat for species such as the eastern cottontail, woodchuck, golden mouse, southeastern shrew, and others that prefer open fields and brushy borders would gradually decrease.

CHAPTER 4

PARK FORMATION

The history of the Great Smoky Mountains is not the simple story of preserving a wilderness, but rather the complex narrative of restoring—and even creating—one.

—Margaret Lynn Brown (2000)

A national park in the Great Smokies was first suggested in 1899. In 1901 James Wilson, secretary of agriculture, stated, "This region in its mountain features, in its forests, and in its climate, stands grandly out as the greatest physiographic feature in the eastern half of our continent." However, very little progress was made until 1924. At a meeting of the Southern Appalachian National Park Committee in Gatlinburg in 1924, Professor H. C. Longwell, head of the geology department at Princeton University, stated: "After all these experiences (seeing mountains from Japan to Greece) I looked down from the top of Mount LeConte. It was wholly unique. In its blending of color, its multiplicity of outline, enveloped into its fairy, ghost-like veil of haze, there is nothing else like it on the face of the earth." In December 1924 the committee recommended that the Great Smoky Mountains would make a fine national park. The committee reported that the Smokies area "surpasses any other region in scenic grandeur" and mentioned these features: "the height of mountains, depth of valleys, ruggedness of the area, and the unexampled variety of its trees, shrubs and plants." Congress gave its approval in 1926 but said that the land for the park must first be obtained by the states in which the park was to be located.

Tennessee and North Carolina began at once to raise money and buy land in the Great Smokies. The legislatures of the two states appropriated some of the

> **Do You Know:**
> When the Great Smoky Mountains National Park was dedicated?
> Which president officially dedicated the park?
> The current size of the park in acres?

The Champion Fibre Mill at Smokemont was the site of major logging operations in North Carolina.

money, as did the city of Knoxville. Individuals, including schoolchildren, also contributed. When this work started, nearly all of the land was privately owned. There were more than 6,600 separate tracts that had to be secured. Approximately one-third of the area was still primeval forest. On some of the remaining acreage there had been only selective cutting of timber. Much of the rest was in varying stages of reforestation, after having been cut over by lumber companies or cleared by mountain farmers. Most of the area (over 85 percent) consisted of large tracts of land owned by 18 lumber and pulpwood companies. Some 1,200 tracts consisted of the small farms of mountain families. The remainder of the land consisted of over 5,000 lots and summer homes.

The first large tract of 76,507 acres was purchased from Colonel W. B. Townsend, owner of the Little River Lumber Company, for $273,557. It was often more difficult to acquire many of the smaller tracts because there was much opposition to giving up family homes, farms, and vacation homes. Numerous mountain farms had been handed down from generation to generation, with the result that often there was a strong desire to remain at the old home place. Many, however, gladly sold their land. Thornborough (1962) stated, "One farmer, after small returns from much labor declared, as he stood with us on the high ridge behind his mountain holdings, gazing out on range after range of mountains, 'I reckon a park is about all this land is fit for.'" Others sold their land but arranged to remain in their old homes for a number of years. Some were given half the value of their property and a lease that allowed them to live there for the rest of their lives.

The Walker sisters (left to right): Martha, Polly, Margaret, Louisa, and Hettie. Circa 1933–35.

One such family involved six sisters. The Walker sisters (Margaret, Polly, Martha, Nancy, Louisa, and Hettie) had been born and raised in Little Greenbrier. Their three-room, two-storied log house was built in the 1870s by their father, John Walker, a Civil War veteran. The walls were papered with newspapers and magazines. Their land was appraised in March 1939 for $5,446 and again in September 1939 for $4,428. The sisters' first asking price was $7,000, after which they made compromise offers of $6,500 and $5,500, neither of which proved acceptable. A lifetime lease was a primary part of each compromise offered by the sisters. Finally, in late 1940, faced with condemnation, they accepted $4,750 for their land, provided they were "allowed to reserve a life estate and the use of the land for and during the life of the five sisters." On January 22, 1941, ownership of the Walker sisters' land (122.8 acres) passed to the United States, but the sisters remained until their deaths. Even though they were allowed to live on their land after it was acquired by the park, they had to adhere to park regulations and modify some of their traditional activities such as hunting and fishing, wood cutting, herb gathering, and livestock grazing. For many years, visitors could drive to their cabin and see them working in the fields and weaving their own clothes. In 1962 I had the privilege of meeting the only remaining sister, Louisa and her brother, Dan. Louisa enjoyed writing poetry about her beloved mountains and home and selling them to visitors. I purchased one of her poems during my visit in 1962 and have kept it in a scrapbook for the past 45 years. I have never seen this poem published in any of the books about the Walker sisters, but it is certainly appropriate for this natural history book. It reads as follows:

Louisa Walker, the last surviving Walker sister.

Little Green Brier Valley

This is a beautiful place
And I love it well
This little Green Brier Valley
In which I dwell.

There is a beauty around us
Wherever we look
Even in the willow
That hangs over the brook.

I love to live
At the foot of the mountain
And drink the pure water
From the sparkling fountain.

Our orchard too
Is a beautiful thing
When the fruit trees bloom
In the early spring.

There the bee's hum
And the dove's lovely coo
And the note of the whippoorwill
Is often heard there too.

The mountains that surround us
Is a beautiful scene
In spring time and summer
When the trees are green.

Then in autumn
More beauty unfolds
When the leaves change
Their color to red and gold.

Some of spring time, summer
And autumns beauty
I have told to you
I must now speak of winter and
    its evergreens too.

There is beauty in the hemlock tree
With its branches spread
In the holly all covered
With berries red.

There is a beauty too
In the mistletoe
With its bead-like berries
All aglow.

The turkey berry vine
And every green fern
Dots the road side
Wherever we turn.

It is a beautiful place
Indeed you know
When the ground is all covered
Up in snow.

And another thing that makes
Our valley look nice
Is the sunshine bright
On the sparkling ice.

                              Composed by
                              Louisa Walker
                              Route 7
                              Sevierville, Tennessee

Upon Louisa's death in July 1964, the home and property reverted to the National Park Service, which maintains them for park visitors. While the land immediately surrounding the home, corncrib, and springhouse is also being maintained by the park service, the fields and clearings are reverting to forest. My wife and I last hiked to the Walker sisters' homesite in November 2005. While walking around the area, I found it increasingly difficult to visualize the former extent of the original homesite with its fields and orchards than had been the case on former visits. After some 50 years, evolutionary succession has obliterated most evidence of human occupation.

Steve Woody's place in Cataloochee Valley, a typical mountain home, in 1938.

As I hike through other areas of the Smokies, I sometimes find apple trees, boxwoods, daffodils, periwinkle, and other "domestic" plants. These are persistent remnants of the homesites that dotted these hillsides many years ago. Now surrounded by the wild species of the forest, they are often all that remain to indicate that some human soul once called that place home.

By 1931 the states had obtained enough land so that the National Park Service could begin to develop the park. A program of protection and general improvement was started. A total of nearly $5 million was raised by the two states. But this was only about half of the amount needed. John D. Rockefeller Jr. matched the amount raised by the states as a tribute to his mother, Laura Spelman Rockefeller. The U.S. government also provided some help. On September 2, 1940, Great Smoky Mountains National Park, comprising approximately 463,000 acres, was officially dedicated. A monument at Newfound Gap reads: "This Park Was Given One-half By The Peoples And States Of North Carolina And Tennessee And By The United States

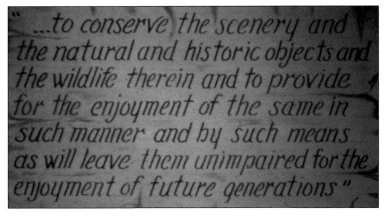

National Park Service mandate.

of America And One-half In Memory Of Laura Spelman Rockefeller By The Laura Spelman Rockefeller Memorial Founded By Her Husband John D. Rockefeller."

In the huge acquisition program to create the park, there was only one outright donation of real estate (Lix, 1958). That was the 102.3-acre Voorheis tract known as Twin Creek Orchard, located between Gatlinburg and Cherokee Orchard on the slopes of Mount LeConte. The donation, made in 1933, contained a provision for Mr. and Mrs. Voorheis to continue living there during their lifetimes. Mr. Voorheis died in 1934. On March 3, 1952, Mrs. Voorheis relinquished her interest in the estate to the National Park Service. The tract had five houses, a large barn, and several smaller buildings. Shortly after the purchase, the superintendent moved there

Plaque at Newfound Gap commemorating the park's dedication.

President Franklin Roosevelt speaking at the park dedication at Newfound Gap on September 2, 1940.

from the residential area near park headquarters. The Twin Creeks site has served as housing for park personnel and more recently as the center for scientific research in the park. It is the site of the park's science center, which was completed in 2007 (see chapter 12).

The park is continually changing in size; for the most part it is growing larger (Table 4.1). In March 1948 the addition of 44,170 North Carolina acres received from TVA in connection with the Fontana Dam project, along with the acquisition of several newly acquired small tracts, brought the total acreage in the park up to 490,948 acres. In 1983 the park purchased a 2,169.70-acre inholding near Fontana Lake from Cities Services Oil Company. By 1993 the park had grown to include a total of 520,000 acres, including nearly 9,457.54 acres in the Foothills Parkway, an unfinished two-lane scenic highway which circles—but does not touch—the Tennessee side of the park. The Foothills Parkway was authorized on February 22, 1944, "to provide an appropriate view of the Great Smoky Mountains National Park from the Tennessee side." In January 2005 the Smokies transferred 110.76 acres of submerged National Park land to ALCOA Power Generating Inc. in exchange for 287 acres of dry land bordering the Smokies. In 2004 a rider was added to the Interior Appropriation Bill (H.R 1409) in the U.S. Congress that required the National Park Service to transfer the 141.67-acre Ravensford Tract from Great Smoky Mountains National Park to the Eastern Band of Cherokee Indians. In exchange, the Eastern Band purchased 218 acres of land to be added to the Blue Ridge Parkway.

There continue to be numerous smaller boundary adjustments and donations of lands. As of January 2008 the size of the park stood at 521,257.24 acres.

The first park superintendent, Major J. Ross Eakin, officially began his duties on January 16, 1931. The first headquarters or "park office" was located in the Post Office Building at Maryville, Tennessee. In June 1932 the park office moved to temporary quarters in Gatlinburg behind the Mountain View Hotel, where it remained until the present administration building near Gatlinburg was completed and occupied in February 1940 (Lix, 1958; Thornborough, 1962).

Table 4.1. Acreage of Great Smoky Mountains National Park, 1940–2007.

Year	Area (in acres)
September 1940	463,000.00
January 1, 1950	490,958.59
January 1, 1960	509,181.98
January 1, 1970	514,601.71
January 1, 1980	517,660.08
January 1, 1990	520,003.78
January 1, 2000	520,977.31
January 1, 2008	521,257.24

Data compiled by R. W. Wightman, Great Smoky Mountains National Park.

One may receive a variety of answers when asking when the Great Smoky Mountains National Park was established. President Calvin Coolidge signed the bill authorizing establishment of the park on May 22, 1926. The first superintendent began his duties on January 16, 1931. The official date recognized by the National Park Service is June 15, 1934, the date on which Congress authorized full establishment, for full completion. President Franklin D. Roosevelt dedicated the park on September 2, 1940. Even though the park was officially established in 1934, all logging had not yet ceased. Little River Lumber Company sold its land with a provision that it be allowed to finish logging it. Its last trees were cut in the small valley of Spruce Falls Branch near Tremont in 1938.

CHAPTER 5

FORESTS AND BALDS

A nation that destroys its soils destroys itself. Forests are the lungs of our land, purifying the air and giving fresh strength to our people.

—Franklin Roosevelt (1937)

Introduction

If allowed only one word to justify the Smokies' worthiness as a national park, that word would be "plants." Arthur Stupka, the first park naturalist, stated, "Vegetation to Great Smoky Mountains National Park is what granite domes and waterfalls are to Yosemite, geysers are to Yellowstone, and sculptured pinnacles are to Bryce Canyon National Park."

It is not known when the name Great Smokies originated. However, it does appear on some early maps. Other maps designate the entire boundary line between North Carolina and Tennessee as the Iron Mountains, Alleghany Mountains, and Unaka Mountains. These mountains were first officially referred to as the "Smoky Mountains" in the act of cession passed in 1789 by North Carolina, when that state offered to cede its western lands to the federal government: "Thence along the highest ridge of said mountains to the place where it is called the Great Iron or Smoky Mountains."

Beginning in 1947 Robert Whittaker undertook a comprehensive study of the

> **Do You Know:**
> How many species of plants exist in the park?
> How many species of trees grow in the park?
> How glaciation affected the flora and fauna of the park?
> How the Great Smoky Mountains got their name?
> How much of the park consists of old-growth forest?
> The meaning of the term "bald" and how many of them occur in the park?

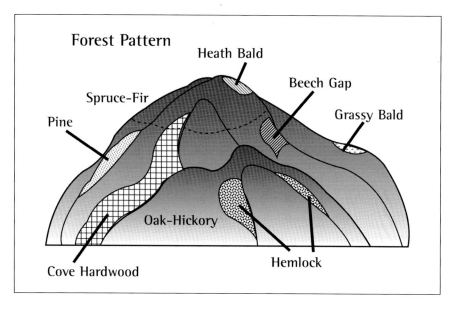
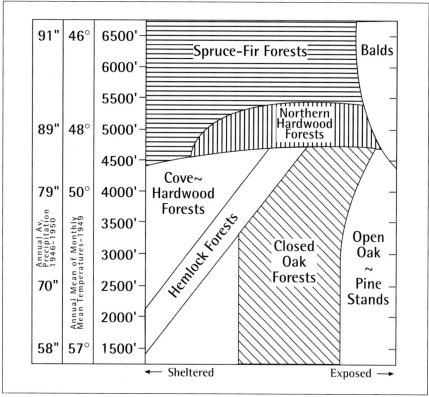

Spatial and elevational relationships of major forest types in Great Smoky Mountains National Park. Based on diagrams by R. H. Whittaker, 1952, 1956. Illustrations by Joey Heath.

vegetation in the Great Smoky Mountains. It was, as Whittaker stated, "an experiment in population analysis of a whole vegetation pattern." He analyzed the relationships of species populations to one another and environmental gradients, and also trends in community composition and structure along environmental gradients. Whittaker classified the climax vegetation into 15 types: cove hardwoods forest, eastern hemlock forest, gray beech forest, red oak–pignut hickory forest, chestnut oak–chestnut forest, chestnut oak–chestnut heath, red oak–chestnut forest, white oak–chestnut forest, Virginia pine forest, pitch pine heath, table mountain pine heath, grassy bald, red spruce forest, Fraser fir forest, and heath bald.

Previous studies by S. A. Cain dealt with heath balds, subalpine forests, cove forests, floristic affinities, the relationship between soil reaction and plant distribution, and the description of a number of vegetation types. W. H. Camp and B. W. Wells reported their findings on the grass balds, and N. H. Russell discussed the beech gaps.

Great Smoky Mountains National Park is probably the most biologically diverse region in all of temperate North America, a feature recognized by its designation as an International Biosphere Reserve in 1976 and by its certification as a UNESCO World Heritage Site in 1983. Almost 95 percent of the park is forested, and about 25 percent of that area has not been disturbed by logging or agriculture. Over a dozen species of trees attain record size in the Smokies, with some being over 20 feet in circumference.

Within the park are found 1,660 species of vascular plants, including 68 ferns, fern allies, and horsetails. Nonvascular plants such as the mosses, liverworts, and hornworts (bryophytes) total 491. The flowering plants include 35 kinds of delicate orchids, 27 kinds of violets, and 58 members of the lily family. There are 158 species of shrubs (114 native, 44 non-native), 32 species of vines (23 native, 9 non-native),

The yellow-fringed orchid can be found at elevations ranging from 2,000 to 4,000 feet. It blooms during July and August. No two flowers have the identical fringe pattern.

In the flowering dogwood, the flowers actually occur in small, compact clusters in the center of a cluster of four showy, whitish bracts, which are usually considered to be the petals. The unique bark is broken into tiny squares.

and 137 species of trees (102 native, 35 non-native). There are more tree species in the Smokies than in any other U.S. national park in North America. Several plant species, including Cain's reed-bent grass and Rugel's ragwort (Rugel's Indian plantain), are not found anywhere else in the world. New species are continually being found (see chapter 10).

Most of the flowering plants burst into bloom before the first day of June, with others following as summer creeps up the slopes. Many of the same flowers that bloom in March at the lower elevations may bloom nearly three months later in the high country. From early spring until late fall the park blazes with color, and in autumn the changing leaves put on another show of color.

Fire pink is usually found on dry, steep banks along roads and trails at elevations between 1,500 and 2,500 feet. The flowers are brilliant scarlet and notched at their tips.

Approximately 2,700 species of fungi occur in the park. One of the easiest to recognize is the turkey tail fungus. Since fungi lack chlorophyll, they cannot produce food from inorganic material.

Caesar's amanita mushroom. Most people never see the mushroom plant but only the fruit it produces—the "mushroom." The plant that produces the fruit consists of a mass of threads, each of which is known as a mycelium. Mycelia secrete enzymes that digest food particles. When the digested material is in solution, it is absorbed by the mycelia and used in the life processes of the fungus.

In addition, the park is home to more than 2,700 species of fungi, 563 species of lichens, and 952 species of algae.

Over 300 species of native vascular plants as well as many of the nonvascular plants, fungi, lichens, and algae are considered rare, meaning they are generally found in small populations, have five or fewer occurrences within the park, or have not been found in recent decades. The park is home to three federally listed threatened (T) and endangered (E) plant species: spreading avens (E), Virginia spiraea (T), and rock gnome lichen (E), the latter being part fungus (see discussion in chapter 7). A total of 76 species of park plants are currently listed as threatened or endangered in the states of Tennessee and North Carolina.

Glaciation

Glaciation during the Pleistocene epoch (500,000 to 20,000 years ago) had an indirect but significant impact on the flora and fauna of the Great Smoky Mountains. During the last Ice Age, some 15,000 to 20,000 years ago, extensive ice sheets formed over eastern Canada and moved south and west into northern portions of the United States. Other glaciers formed over Scandinavia and moved southward over northwestern Europe and over the arctic regions of Eurasia and North America. The ice sheets covering Greenland and Antarctica also grew larger. However, glaciers never extended as far south as the Smoky Mountains. At its closest point, ice extended as far south as the Ohio River. By this time, the effects of weathering (freezing, thawing, erosion) had covered the mountains with rich soil. If the glacier ice had not stopped at the Ohio River, it would have stripped away this

Distribution of glaciers during the Ice Age. Glaciers locked up so much water in the form of ice that sea levels worldwide were lowered by approximately 100 meters. We are now in a period of global warming. Temperatures are increasing, remaining glaciers are receding, and large pieces of ice are breaking free of Antarctica. If this warming trend continues, it will undoubtedly affect the worldwide distribution of plants and animals. From *Vertebrate Biology* by D. W. Linzey (New York: McGraw-Hill, 2001).

soil—as it did in New England, where the advancing glaciers created such features as New York's Finger Lakes.

During the ice ages, it is thought that some of the mountain peaks of the Smokies were probably covered with snow and had tundra vegetation. Some areas may have had a permanent year-round snowpack. The huge boulders, locally known as "graybacks," that form boulder fields on the slopes and in the valleys today are thought to have been broken off from the tops of the mountains by intense freezing-thawing activity. Later, as the glaciers retreated and the climate warmed, boulder creation slowed and now happens infrequently. You can actually find the source of these boulder fields by following them up to their original rock outcrop. The more

"Graybacks" in a boulder field near Chimneys picnic area.

recent rocks that have been broken off are jagged. The rocks that were broken off long ago are more rounded, due to greater exposure over time to weathering by wind and water.

The northeast-to-southwest orientation of the Appalachian Mountains allowed species to migrate southward along the slopes rather than finding the mountains to be a barrier. Thus, Canadian-zone flora gradually seeded southward, and fauna retreating before the rivers of ice established new footholds high in the Smokies, where some still survive. When the ice age ended, the glaciers retreated and many of the northern plant species followed, reseeding northward. As the temperatures rose, others seeded themselves up the slopes to the higher, cooler elevations. As they vacated the lower levels, southern plant species returned to take their place. Many forms of animal life did likewise. Thus, the Great Smoky Mountains have both northern and southern species of plants and animals. Because of the glaciers and the range of elevations, there are 137 kinds of trees now growing in Great Smoky Mountains National Park—more than in any other area this size in the temperate zone of North America and more than in all of Europe. There are species such as the northern water shrew (*Sorex palustris*), the northern flying squirrel (*Glaucomys sabrinus*), and the spruce-fir moss spider (*Microhexura montivaga*) that are near the southernmost limits of their range in the park and survive in disjunct (isolated) populations. Other species such as the smokey shrew (*Sorex fumeus*), southern red-backed vole (*Clethrionomys gapperi*), and rock vole (*Microtus chrotorrhinus*) are also approaching the southern limits of their range in the Great Smokies but are more widespread throughout the park.

Forests and Balds

Forests

In this part of the world, forests are usually the last step in the plant and animal succession that begins with rock colonization and soil formation. In Great Smoky Mountains National Park, we can see various stages in this progression because natural disturbances such as fire and windstorms and humans with their cutting, burning, and livestock grazing have set back the successional clock. Rock, meadow, brush, pioneer forest, and mature forest—each has distinctive plants and animals that fade and are replaced as each stage gives way to the next.

Decaying logs not only release nutrients back into the soil but also provide substrate, food, and shelter for many plants, invertebrates, and vertebrates.

Coarse woody debris (CWD), in the form of downed logs and standing dead trees, offers vital habitat for many organisms and serves an important role in the long-term cycling of nutrients in forests. Results of a study in the park by Dr. Chris Webster of Michigan Technological University suggest that areas with a history of human disturbance may require well over a century to recover coarse woody debris distributions found in areas of primary forest.

Variations in elevation, rainfall, temperature, and geology in these ancient mountains provide ideal habitat for a great variety of tree and shrub species. What is the difference between a shrub and a tree? Trees are, of course, woody plants, but so are so-called shrubs. We often recognize a tree as being larger than a shrub, and usually think of them as having a solitary stem or trunk. Actually the differences between trees and shrubs are purely relative ones. The difference is somewhat similar to asking what is the difference between a pond and a lake or between a hill and a mountain. No real line of demarcation exists between trees and shrubs. Frequently, a species will be merely shrubby in one portion of its range and be quite large and treelike somewhere else. The generally accepted botanical definition is that a tree is a woody plant with a well-defined stem and crown, has a diameter of two inches or more, and attains a height of at least eight feet. This definition includes a number of plants that ordinarily are shrublike, but which, on occasion, become treelike, such as mountain laurel, alternate-leaf dogwood, common alder, witch-hazel, staghorn sumac, several kinds of plums, hawthorns, *Viburnum* sp., mountain stewartia, and devils walkingstick. The rosebay rhododendron and the mountain laurel are usually only shrubs in the mountains of Pennsylvania, yet they may grow to 40 feet in height and assume treelike proportions in the mountains of North Carolina and Tennessee. For example, Stupka (1964) reported a mountain laurel in the park that was about 25 feet tall with one of its numerous trunks measuring 3 feet, 6 inches in circumference. He also recorded a shining or winged sumac with a trunk circumference of 13 inches, a common alder with a 22-inch circumference, and an alternate-leaf dogwood that was 30 feet tall and had a circumference of 12 inches. If these are counted as trees, then there are 137 kinds of trees in the Great Smokies.

New species of trees are still being found in the park. The most recent discovery occurred in 2004 when the stolon-bearing hawthorn (*Crateagus iracunda*) was discovered. The population in the park is the largest known population for this southeastern tree.

The forests are the primary cause of a blue haze that typically hangs over the Smoky Mountains and gives them their name. Horace Kephart, in his book *Our Southern Highlanders*, described the "smoke" as "the dreamy blue haze . . . that ever hovers over the mountains." Through transpiration, the trees give off a volatilized oil known as terpene. These hydrocarbon molecules break down in sunlight and recombine to form molecules that are large enough to refract light and that react with each other and various pollutants to create the smokelike haze.

Many trees in the park predate European settlement of the area. By removing core samples, Will Blozan of the Eastern Native Tree Society has found a 333-year-old yellow birch, a 341-year-old pignut hickory, a 351-year-old white pine, a 359-year-old white oak, a 395-year-old red spruce, a 400-year-old shortleaf pine, a 412-year-old chestnut oak, a 500-year-old tuliptree, a 517-year-old eastern hemlock, and the oldest, a 578-year-old black gum. All ages are as of 2008, and all trees are assumed to be alive.

Five major communities of tree associations (forest types) occur. They are the spruce-fir forest, northern hardwood forest, cove hardwood forest, pine-oak forest, and hemlock forest. Some authors recognize six (closed oak forest; Stupka, 1960) and even seven (closed oak forest and lowland streambank forest; Stevenson, 1967). These forest types have developed as a result of such factors as temperature

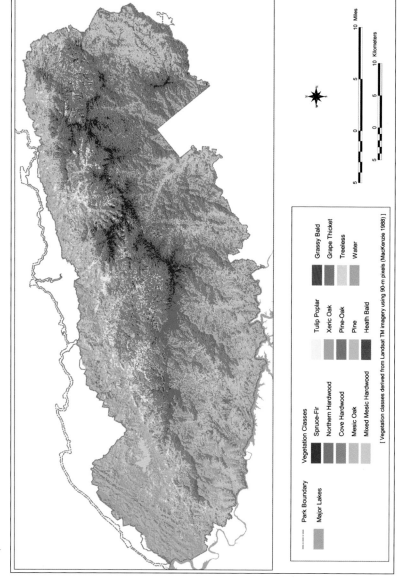

Vegetation in the Great Smoky Mountains National Park.

and length of growing season (both of which are correlated with altitude) and by moisture and soil depth (both of which are correlated with topographic position). As elevation increases within the park, temperature decreases and precipitation increases. As noted earlier, every 1,000-foot rise in elevation results in a decrease of two to three degrees Fahrenheit and is the equivalent of moving 250 miles northward. The cool, wet conditions on the summit of Clingmans Dome actually makes the spruce-fir forest that grows there a coniferous rainforest.

Spruce-Fir Forest

The main components of the spruce-fir forest are red spruce and Fraser fir. Fraser fir is easily distinguished from red spruce by its upright cones (no other tree in the park has upright cones) and by its blunt aromatic needles, which are green above and lined with gray below. Bark blisters filled with clear rosin are present in the thin bark of many Fraser fir trees. Early settlers imagined these blisters to be filled with rosin "milk," and so called them "she-balsams." The red spruce, which grows to a greater height and diameter, has pendulant (downward-hanging) cones and sharp-pointed needles that are the same shade of green above and below. The bark has no rosin blisters; as a result, the settlers called these trees "he-balsams." Since these two kinds of conifers often grow in mixed stands, they evidently assumed that one, the fir, was the female, while the other, the spruce, was the male.

This forest caps the park's highest elevations and has been called the finest forest of red spruce in the United States. Although most areas above 4,500 feet support some elements of this forest, it is best developed above 5,500 feet. The spruce-fir forest extends down to about 4,000 feet on the Tennessee side of the park, but because of more extensive logging operations on the North Carolina side, spruce and fir are rare below Newfound Gap (5,000 ft.). In terms of climate, the spruce-fir forest has a climate similar to areas such as Maine and Quebec, Canada. It is sometimes referred to as an island of Nova Scotian vegetation floating

Water quality and aquatic life in mountain streams are directly affected by the spruce-fir forest since it is the dominant forest at their headwaters.

Fraser fir is the only tree in the park with upright cones. The seeds are a favorite of red squirrels.

Fraser fir (left) with its flat, fragrant, and blunt-tipped needles; red spruce (right) with its square, sharp-pointed needles. Photo by Steve Bohleber.

atop a sea of Appalachian hardwoods. I have traveled and hiked in Maine and Nova Scotia and can personally attest to the similarities.

An almost unbroken forest of spruce and fir extends the length of the park along the Tennessee–North Carolina boundary from the vicinity of Double Springs Gap on the western slope of Clingmans Dome to the vicinity of Cosby Knob near the northeastern corner of the park. The only interruption occurs near Charlies Bunion, an area devastated by the great fire of 1925. The most extensive stand on the Tennessee side of the park occurs on Mount LeConte, the third-highest peak in the Smokies, while the most extensive stand on the North Carolina side extends southward from Mount Guyot (second-highest peak) between Hughes Ridge and Balsam Mountain. Tree associations include American mountain-ash, beech, yellow buckeye, mountain maple, pin cherry, black cherry, Allegheny serviceberry, and yellow birch. Above 6,000 feet elevation, the only trees occasionally associated with red spruce and Fraser fir are American mountain-ash, mountain maple, pin cherry, and yellow birch. All species that exist in these high-elevation forests must be able to endure cold temperatures, short growing seasons, and heavy snow and ice.

In 1990, 36 research plots were established in Fraser fir stands on five peaks in the park, with the objectives of determining the status of the overstory, factors limiting fir regeneration, and characterization of regeneration patterns. Data were collected from these plots in the year 2000 to examine changes during the previous decade. In addition, surviving mature fir stands were delineated; stands of over one hectare were found at Mount Guyot and Old Black, with smaller stands on Mount Sterling, Mount LeConte, Mount Chapman, and Big Cataloochee.

In many areas, the dense growth of trees prevents the growth of shrubs and other understory plants. Their tall dark trunks block most of the sunlight and keep

Spores on the underside of the oblong, leathery, blunt-tipped leaflets of the polypody fern.

The small whitish flowers with pink stripes, together with the shamrock-like leaves, make it easy to identify American wood-sorrel, an abundant ground cover in the spruce-fir forest.

The painted trillium is found on moist shaded slopes at elevations ranging from 3,000 to 6,500 feet. The name "painted" comes from a pink "V" at the base of the white petals. Photo by Steve Bohleber.

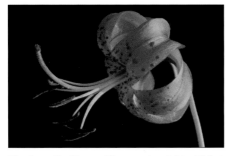

The large Turk's-cap lily may stand 6 to 10 feet tall. It can be seen along the Clingmans Dome Road and at other high elevation sites.

Pink turtlehead is a fairly abundant plant at higher elevations. The shape of the flower suggests its name. When one presses the side of the flower, the "turtle's" mouth opens.

the mossy forest floor in a state of perpetual twilight. The thick layer of needles takes years to decompose due to low temperatures and acidic conditions. In other areas, shrubs such as blueberry, Catawba, rosebay, and Carolina rhododendrons ("rhododendron" is derived from the Greek and means "rose tree"), hobblebush, round-leaved currant, red elderberry, bush-honeysuckle, thornless blackberry, and wild raisin may be found. High-elevation ferns include common polypody fern, hayscented fern, lady fern, and woodfern. The most conspicuous spring-blooming herbs include American woodsorrel, bluet, erect trillium (white and purple forms), painted trillium, pallid violet, spring beauty, and yellow bead lily. Common summer-blooming herbs are goldenrod, Indian pipe, pink turtlehead, Turk's cap lily, and white wood aster.

Hobblebush is a prominent shrub of cool, damp woodlands above 3,000 feet. When low arching branches touch the ground, they often take root. From such rooting points, new shoots develop. The tangle of arching and rooting branches produces a thicketlike growth that can "hobble" those who try to walk through it. Clusters of small white flowers appear in late April and early May, followed in early fall by brilliant crimson fruits. The large, round leaves change from green to variegated hues and reach their peak of autumnal leaf color in late September.

In the transition zones, low- and middle-elevation species of birds are replaced by high-elevation species. For example, the Carolina chickadee is replaced by the black-capped chickadee, the wood thrush by the veery, the common crow by the common raven, and the screech owl by the saw-whet owl.

Five small northern birds reach the southernmost limit of their breeding range in the spruce-fir forest: the golden-crowned kinglet, winter wren, brown creeper, black-capped chickadee, and red-breasted nuthatch. During the summer, individuals of these species usually live alone or as breeding pairs. Later in the year, however, they often gather in mixed-species groups and forage through the woods together during fall and winter.

Since the mid-1980s, the insect known as the balsam woolly adelgid has killed 95 percent of the mature Fraser firs. Accidentally introduced from Europe, the tragedy of this pest threatens the fate of the entire forest type (see chapter 11). The boreal forests live just above the ethereal elevation of air able to support the colder climate they need. As society continues to spew hydrocarbons into the air and global warming continues, this line will inexorably rise, accelerating the rate at which the Smokies' firs, spruces, and ferns are pushed skyward into oblivion. Other environmental pressures, including acidic deposition and ozone, present further threats (see chapter 11).

Spruce-fir forest as seen from Clingmans Dome Road in 1979, before the effects of the balsam woolly adelgid were widely noticed.

Effects of the balsam woolly adelgid as seen from Clingmans Dome, 2001. Photo by Steve Bohleber.

Northern Hardwood Forest

Northern hardwood forests occur mostly above 4,500 feet and are dominated by yellow birch and American beech. These forests resemble those throughout much of New England, New York, Pennsylvania, and southern Ontario. In many cases, they are almost surrounded by red spruce and Fraser fir. Some northern hardwood forest trees such as black cherry, eastern hemlock, and sugar maple reach their uppermost limits at or near 5,000 feet elevation. Others such as Allegheny serviceberry, American beech, red maple, striped maple, and yellow buckeye reach their uppermost limits by 6,000 feet. Mountain maple, pin cherry, and yellow birch may reach the highest summits of the Smokies where spruce and fir are dominant. The northern hardwood forest, specifically sugar

The three-lobed leaves of the striped or "goosefoot" maple. When young, the green bark has white stripes. Photo by Steve Bohleber.

maples, produces the most brilliant fall color.

A relatively common tree up to 5,800 feet in this forest is striped maple. Striped maples are never big trees; they flourish in the understory beneath the canopy of taller trees. The branches and young trunks are boldly marked with thin vertical stripes that range from dark green to chalky white. The large three-lobed leaves resemble the foot of a goose; hence, one of the common names for this tree is "goosefoot" maple.

American beech grows in pure stands in many of the gaps along the high mountain crest and is widespread in the park's lower-elevation forest as well. These trees are distinguished by their smooth gray bark, dark shiny leaves, and a tendency for their withered leaves to cling to the winter branches. Beneath many beech

Striped maple with snail. Photo by Steve Bohleber.

trees is a drab plant of the forest floor that lacks food-producing capacity of its own. Known as beechdrops, this plant parasitizes the roots of the beech. Always look for beechdrops under beech trees.

A limited number of shrubs occur in the northern hardwood forest. These include Catawba and rosebay rhododendrons, dog-hobble, hydrangea, and smooth blackberry. A large variety of herbaceous plants also occur, including American woodsorrel, common fawn-lily, creeping bluet, crinkleroot, fringed phacelia, great starwort, trilliums, violets, Virginia spring-beauty, and yellow bead lily.

The drooping branches of dog-hobble with their thick, leathery, and evergreen leaves form dense thickets in damp places within shady forests. Dog-hobble is found in northern hardwood, cove hardwood, and hemlock forests. The curious name of dog-hobble

Blossom of the Catawba rhododendron. Although normally 8 to 12 feet high, the Catawba rhododendron occasionally grows to the size of a small tree. The national champion with a girth of 15 inches and a height of 25 feet grows along the Baxter Creek Trail in the North Carolina portion of the park.

Hiker in a bed of fringed phacelia near Indian Gap. From a distance, the beds resemble patches of snow.

Fringed phacelia (close-up).

apparently comes from the ability of these thick growths to impede the progress of hunting dogs in their pursuit of black bears which, because of their bulk and strength, can readily move through these thickets. Dog-hobble blooms from late April to June. During autumn, clusters of flower buds form in the angles of the uppermost leaves. As winter advances, these clusters turn deep red, while some of the leaves turn to a maroon, bronze, or copper—a touch of color in the winter woods.

The bell-shaped, strongly scented flowers of dog-hobble hang in clusters and will eventually produce dry, brown seed capsules. Dog-hobble rarely grows more than three feet high.

British soldier lichen, also known as scarlet-crested cladonia (*Cladonia cristatella*), grows on the forest floor or from the moss-covered surface of fallen logs. The name comes from the bright red fruiting cups that reminded early wanderers in the woods of the red uniforms of Revolutionary-period British soldiers. It is relatively pollution-tolerant and often grows where people live. Lichens are not single plants. They are autotrophic (having the ability to make their own food by photosynthesis) organisms composed of a sac or club fungus and a photosynthetic unicellular organism (either a cyanobacterium or an alga). They form a symbiotic, mutualistic relationship. The fungus anchors the organism to a substrate such as a rock or tree trunk, while the photosynthetic cyanobacteria

British soldier lichen.

or alga synthesize carbohydrates. Lichens get minerals from airborne particles and are highly vulnerable to air pollution. They can colonize places that are too hostile for most organisms, with their metabolic products helping form new soil or enrich whatever soil is already present. As conditions improve, other species move in and typically replace the pioneers.

Cove Hardwood Forest

These forests occupy the valleys between the mountain ridges. They occur in sheltered situations and on northern-facing slopes at low and middle elevations (below 4,500 feet) where soil reaches a considerable depth. The cove hardwood forest is the Smokies' most diverse ecosystem. It develops in areas with warm temperatures, a long growing season, and plentiful rainfall. The deep, moist soils support a luxuriant carpet of herbaceous plants, the richest of all forest types in the southern Appalachians. Dominant trees are basswood, eastern hemlock, silverbell, sugar maple, yellow birch, yellow buckeye, tuliptree (yellow-poplar), and, in former years, American chestnut. American beech is important in some stands. Together, these eight trees constitute 80–90 percent of the canopy of cove hardwood forests. Subdominant tree species include red maple, black cherry, Fraser magnolia, white ash, striped maple, ironwood, hop-hornbeam, witch-hazel, and shagbark hickory. Many of these trees grow to record or near-record proportions in the park. Some canopy trees reach heights of 125

The large tuliptree blossoms are often found on trails in May. These straight-growing trees have light gray bark and are one of the most common trees below 4,000 feet elevation.

feet or more and may have trunks six or more feet in diameter. Most tree canopies do not begin until they reach 75 to 100 feet above the forest floor. These forests are the ones that produce spectacular fall coloration. The Sugarlands, Sugarland Mountain, Maple Sugar Gap, Sugar Orchard Branch, and other park places were all named for the large sugar maples that once grew there and provided the sweet sap used in the making of maple syrup and sugar.

Tapping sugar maples was once a fairly common practice in the Smokies. Native Americans used maple sap and sugar to season meats and grains and to make candy and beverages. During the 1800s and early 1900s, many mountain farm families maintained areas in the forest called "sugar camps" or "sugar bushes" for the production of syrup and sugar. Many Smoky Mountain residents described the best time to tap sugar maples as "after the first snow of spring" and "when the strong, warm winds roar down from the mountains." The tapping season could last from two to eight weeks.

Each family might have several dozen sugar maples that had been grooved and tapped to produce sap. Wooden troughs ran from the trees to central buckets or barrels for efficient collection. Family members then carried the sap in buckets to a shed that housed a stone furnace and large metal evaporator pan. Maple sap had to be cooked down for several hours to produce syrup. It generally took 30 to 40 gallons of sap to produce one gallon of syrup. Each healthy sugar maple tree could be counted on to produce between five and 40 gallons of sap. Even more boiling and processing was necessary to make maple sugar.

Maple syrup and sugar were commodities that farm families could consume themselves or trade at a country store for cash or merchandise. In Sevier County, records show farmers produced 38,455 gallons of maple syrup in 1859. "Sugaring" in the region declined sharply in the 1900s, presumably due to commercial logging and easier access to other forms of sugar.

A recent study suggests that leachate from American chestnut leaf litter could have suppressed germination and growth of competing shrub and tree species such as eastern hemlock and rhododendron. Such allelopathy (the chemical influence of one living plant on another, usually in a negative connotation) may have served as a mechanism whereby American chestnut controlled vegetative composition and dominated eastern forests. Current vegetative composition in southern Appalachian forests may be partly attributable to the disappearance of American chestnut as an allelopathic influence.

Twice a week as part of my naturalist duties at Cosby Campground I would lead a car caravan to Indian Camp Creek and a hike into the virgin timber of Albright Grove to see the park's largest known tree, a tuliptree (yellow-poplar)—135 feet high and over 25 feet in circumference. Everyone was amazed at the sizes of the yellow-poplars in this old-growth forest. I always carried a tape measure so they could personally determine the circumference for themselves. Many had their pictures taken as they encircled the giant tree while holding hands. Will Blozan, of the

The cove hardwood forest is the richest of all forest types in the southern Appalachians. The soil is deep and moist and supports the most diverse ecosystem in the park.

Eastern Native Tree Society, who informed me that this tree fell in 1997, counted approximately 400 annual rings (400 years old) on a portion of the trunk about 50 feet above the base. I had last seen the tree on November 2, 1996, when my wife and I hiked to the grove to show the tree to my wife's two sisters. If the definition of "largest" is the tree that contains the most wood but is not necessarily the largest in circumference, then a tuliptree on Sag Branch in Cataloochee (4,100 cubic feet) is currently the largest tree documented in the park (Blozan, 2006).

Ironwood is a common low-elevation tree that grows along stream banks. Its distinctive trunk looks like the flexed muscles of a bodybuilder, hence its common name of "musclewood." The wood has the distinction of being the densest wood of any tree in North America.

Another common tree of the cove hardwood forest is the Fraser magnolia. Its "eared" leaves arranged in superficial whorls and its large cream-colored flowers make good field characteristics for separating this species from the other two native magnolias in the park. The cucumbertree and the umbrella magnolia have leaves that are pointed at both ends. The leaves of the Fraser magnolia, which often reach a length of nearly two feet, are among the largest of any tree in the park. The cucumber-like fruits begin to color in July; by early August they are bright red. As the individual pods of the fruit ripen, they dry and split open, but the seeds they contain remain attached by slender threads. Perhaps the dangling seeds make them more obvious to the birds upon which the tree depends for the dispersal of its seeds to new locations. A record tree with a circumference of over nine feet grows

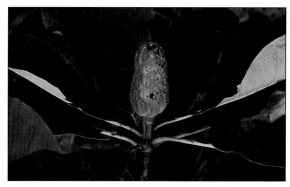

The large red fruit (seed pods) of the umbrella magnolia are conspicuous and are often found along trails in late summer and fall. The large leaves (18–24 inches) and the pointed base of the leaves serve to differentiate this tree from the Fraser magnolia.

near Anthony Creek on the trail to Russell Field at an elevation of about 2,700 feet. The park contains a number of Fraser magnolias over eight feet in circumference and at least 80 feet in height.

Witch-hazel is an irregularly shaped shrub or small tree that usually has several small trunks. It is common on moist, shaded slopes throughout the low and middle elevations of the park. The leaves are coarsely toothed, dark, and lustrous, and often asymmetrical. Just as it is losing its leaves in October, the witch-hazel bursts into bloom. Small clusters of curious bright yellow flowers, each with four twisted ribbon petals, offer a brilliant accent to the fading autumnal woods. In most plants, fertilization comes right after pollination, but witch-hazel delays fertilization until warm weather in the spring. Woody, fuzzy, knobby seed capsules develop through

The flowers of witch hazel appear in the fall just as the plant is losing its leaves. The leaves, which have uneven bases and wavy edges, are nearly as broad as they are long. The fruit is a capsule that may remain on the tree for a year.

the summer from last year's flowers. As they mature, they shrink and split open at the top. Continued shrinking applies pressure on the two black seeds within. Suddenly, and with considerable force, the seeds are ejected up to 10 to 15 feet through the air. The seeds germinate 18 to 24 months after they are ejected.

Have you ever seen a tree propped up on stiltlike roots and wondered how and why that occurred? Yellow birch seeds can germinate on the abundant moss-covered logs of the Smokies woodlands. Roots grow rapidly down around the log to reach the soil. As the tree grows, the log originally supporting it usually decays, and the tree is left propped up on stiltlike roots. Some host logs decay very slowly. Will Blozan has cored a yellow birch over 90 years old that was still on the host log.

The nurse log on which the yellow birch seed germinated has decayed, leaving this yellow birch propped up on stiltlike roots.

One of the main shrubs in cove hardwood forests is the rosebay rhododendron. It is common throughout most of the park except in the spruce-fir forest, where it is rare. This species grows best in the shade of forested ravines and slopes along streams where it forms dense, nearly impenetrable thickets. At low elevations, it blooms in June. Higher up the slopes, flowering occurs in July or even August. The waxy white or pinkish flowers are marked with greenish-yellow spots on the upper lobe. They attract large insects such as bumblebees and hawk-moths as pollinators. The hovering flight and long tongue of the hawk-moth are special adaptations that allow it to reach the nectar deep within the flower tube. This evergreen plant affords food and cover for many wildlife species in winter weather. In

During a drought or when temperatures drop below freezing, the evergreen leaves of rhododendron roll into a coil and droop.

subfreezing temperatures and during severe droughts, the thick, dark evergreen leaves roll into a tight coil and droop.

Another fairly common shrub at lower and middle elevations is heart's-a-bustin'. It favors the rich and moist soils of cove hardwood forests. The inconspicuous green or greenish-purple flowers appear in May. The flowers ripen into spiny or warty coral-pink pods. In early autumn, these burst open to reveal the brilliant orange-red seeds within. The resemblance of the unopened pod to a strawberry gives the plant another of its common names: strawberry-bush.

Spicebush is one of the earliest shrubs to come into flower. Yellow flowers appear in early March, well before the leaves. The green fruits turn bright, shiny red in August and are eaten by many birds. Spicebush prefers moist soils along streams at lower elevations. When the leaves and twigs are bruised, they emit the clean, spicy odor that gives the plant its name.

American mistletoe seems to be most common on oaks (especially scarlet oaks) and black gums. It is a parasite, drawing all of its water and nutrients from the tree on which it grows. The mineral and water resources of the host tree are tapped by rootlike growths (called haustoria) that penetrate the tissues of the tree. Birds eagerly eat the mistletoe's white and waxy berries. The berries pass through the bird's digestive tract unharmed and are scattered in droppings throughout the woods. When dropped on a suitable branch, they sprout. Seeds are also spread by birds rubbing their beaks on a twig trying to remove the sticky seeds. A compound identified as "viscin" that coats the seeds is a sticky gum that hardens and secures the seed to its new host.

Trout lily, one of the earliest blooming wildflowers, is widely distributed at lower elevations and may be found as high as 6,000 feet. Although a lily, this plant is often called "dog-tooth violet."

Indian pipe is a saprophyte that derives its nourishment from dead or decaying organic matter. The leaves are rudimentary and scalelike.

Many spring-blooming herbs, including the trout lily, are found here. Leaves and flowers must be produced in the brief time before the leaves of the overhead trees shade out the forest floor. Trout lilies do not bloom every spring. Many patches of the speckled leaves without flowers can be found. Curiously, all of these nonflowering plants have but a single leaf. When a trout lily blooms, it always produces two leaves. Following its burst of spring blooming, the trout lily's food reserves are so depleted that the plant must rest for several years. During this time, no flowers and only single leaves are produced each spring. After several years, enough energy has been stored in the bulb to support another spring flowering. As with most spring wildflowers, the trout lily's rapid burst of spring growth is fueled by energy stored in an underground bulb.

The white, waxy appearance of Indian pipe gives rise to another of the plant's common names: ghost flower. When young, Indian pipes are usually pure white (sometimes pinkish-white). Their flowers nod. As they mature, the flowers lift upward and the plant becomes black and brittle. Indian pipes lack the leaves and green pigments (chlorophyll) of most other flowering plants. Their leaves are reduced to scalelike structures. As a result, they cannot make their own food. They live instead on the dead and decaying remains of other plants. The Indian pipe's root system is a compact ball of small rootlets tightly interwoven with fungal strands. Experiments have shown that the presence of the fungus is necessary if the plant is to gather nutrients from the surrounding soil.

Doll's eyes, also known as white baneberry, is a plant of rich woods. It is usually found at middle elevations in the park, often within the cove hardwood forest.

The white fruits of doll's eyes mature in August or September.

Jack-in-the-pulpit, also known as Indian turnip because of its deeply buried, turnip-shaped, solid, bulb-like stem base, is widely distributed in the lower elevations. Native Americans and other have used this plant to treat asthma, bronchitis, boils, coughs, laryngitis, headaches, and swelling from snakebites.

The fruit is a glossy white berry with a dark black or purplish "eye." Its resemblance to the eye of an old-fashioned doll is responsible for the plant's common name. The fruits are always borne in clusters on thick reddish stalks.

Jack-in-the-pulpit is found in damp woods and may appear as early as March. It comprises a purple and green mottled hood or spathe (the pulpit) and knob-shaped flower cluster or spadix (the Jack) consisting of both male and female flowers. The spathe is vase-shaped and tapers to a delicate point. The respiration of the spadix warms the surrounding air, and the enveloping spathe helps to insulate the plant, creating a microclimate of constant warmth. Flowering occurs from March through June. The flowers develop female parts first, beginning at the top of the spadix. By the time female flowers emerge at the bottom of the spadix, male parts have developed at the top. An opening of the spathe allows for the entry and pollination by some of the earliest flying insects. The fruit, a cluster of red berries on the spadix, appears late in the summer through fall and is relished by birds and small mammals. Jack-in-the-pulpit is a member of the arum family and contains needlelike crystals of calcium oxalate that produce a burning sensation if the plant parts are eaten raw. Drying eliminates this unpleasant characteristic. Native Americans gathered the fleshy taproots as a vegetable.

Wild ginger is a most unusual plant. It grows abundantly in rich soils on rocky hillsides up to about 3,000 feet elevation. The paired kidney-shaped leaves are three to six inches in diameter and are usually located close to the ground. The single brownish-purple bloom is hidden beneath the leaves, frequently lying on the ground and often somewhat concealed by the leaf litter on the forest floor. The calyx is cuplike and has three slender prongs extending outward from its edge. It is actually a flower without petals. Another unique feature of this strange plant is its roots. When a root is broken off, it emits a very pungent odor—the reason it is known as wild ginger. Early settlers made from these roots a brew that was believed to be effective in the treatment of colic and certain other types of stomach distress. Another member of the same family (Asaraceae), the little brown jug, possesses thick, evergreen, arrow-shaped leaves and a small, purplish-brown, fleshy, jug-shaped calyx (a flower without petals) near the ground.

A third member of this same family is the Dutchman's-pipe vine. The rope-like woody vine may climb 100 feet into the tops of trees. The vine possesses large heart-shaped leaves and brownish-purple blooms that somewhat resemble the form of an S-shaped Dutchman's pipe. Surrounding the boll of the "pipe" is an expanded flange of mottled color. The narrow throat of the flower is yellow. These strange blossoms emit an odor attractive to small flies. Once these flies enter the bloom, they find themselves trapped for a time in its expanded base. They become well dusted with pollen, eventually escape, and enter other Dutchman's-pipe flowers where their pollen is rubbed off upon the flowers' stigmas, thus completing the process of cross-pollination.

One of the most unique ferns in the park is the walking fern. This evergreen fern is not uncommon but is often overlooked. It clings to shaded ledges of sandstone or limestone. I have found it most recently along Roaring Fork and along the trail to Baskins Creek Falls. It is more common, however, in

A fern that walks. The long, narrow, fine-pointed, arching, evergreen leaves of the walking fern often give rise to new plants whenever their tips touch the ground. Photo by Steve Bohleber.

The bird's-nest fungus resembles a miniature bird's nest complete with "eggs." The eggs (periodioles) are actually spore packages, which rupture to release the reproductive cells. This fungus is sometimes found in great profusion on old sticks, wooden bridges, fence rails, and other unused wood.

the western section of the park. The walking fern always grows from the mossy surface of a moist boulder or rock ledge; it is never rooted in the soil of the forest floor. Where the tip of a leaf touches the ground, a new plant often develops. Thus, the fern seems to "walk" across the rocky face on which it lives.

Many types of fungi are found in the park (see chapter 10). Some, such as bracket fungi, coral fungi, morels, and earthstars, are generally large enough to be easily visible. However, it takes a keen eye to discover one of the most unusual fungi, the fluted bird's-nest fungus (*Cyathus striatus*). It grows on fallen logs and pieces of bark and resembles a miniature bird's nest. Each fungus consists of a vaselike, flaring cone with ribbed sides (the nest) and a number of "eggs" resting within. These "eggs" are actually spore cases, called *periodioles*, and it is the manner of their dispersal by drops of rain falling into the fungal cups that is so unique. When a raindrop falls into one of these cups, the periodioles are splashed as far as seven feet away. The outer wall of the spore case then decays or is eaten away by insects, and the spores within are exposed. They germinate and produce new crops of bird's-nest fungi.

Pine-Oak Forest

Dry, exposed, usually rocky slopes and ridges contain stands of oaks and pines. Despite plentiful amounts of rain, these excessively drained slopes dry out often, and lightning-caused fire was a regular part of these forest communities. Four

A dwarf creeping shrub, trailing arbutus is fairly common in sandy, acid soils up to elevations of about 4,000 feet. It has coarse, hairy, evergreen leaves. The small pink or white tubular flowers appear from early March through May and have a spicy fragrance.

kinds of oaks and several kinds of pines dominate these forests. Since the trees do not normally form a closed canopy, the shrub layer may be quite dense and is often dominated by the evergreen mountain laurel. Oaks include black, chestnut, scarlet, and white oak. Pitch, table-mountain, and Virginia pines are most plentiful on the driest sites, while Allegheny serviceberry, blackgum, black locust, eastern white pine, sassafras, shortleaf pine, and sourwood may also be present. Evidence of the former presence of American chestnut exists in the form of basal sprouts. Flame azalea, rosebay rhododendron, and wild hydrangea are also found here. Asters, blueberries, wintergreen, eastern bracken, galax, huckleberries, pussytoes, trailing arbutus, and pipsissewa are common.

Black oaks sometimes bear ping-pong ball–like growths among their leaves. Because of their resemblance to small apples, they are often called "oak apples." Oak apples are galls—characteristic outgrowths of plant tissue made in response to some irritant introduced into the plant. Oak apples are the work of a tiny wasp that lays its eggs in the buds and soft tissue of young leaves. When the eggs hatch, chemicals given off by the larvae disrupt the leaf's normal growth, stimulating it to produce the gall instead. The larvae feed on the gall tissues, which surround and protect them. When you break open oak apples, you often find the tiny grublike larva still inside. On other galls, a small exit hole indicates that the adult wasp has already emerged.

Hard mast, such as acorns, hickory nuts, beech nuts, and walnuts, is probably the most important fall food for wildlife in the park. Annual variations in hard mast production affect food habits, movements, habitat preference, reproduction, and therefore density of black bears. Hard mast also is an important fall food for other wildlife species including white-tailed deer, wild turkey, chipmunks, squirrels, and wild hogs. Hard mast surveys are used to collect baseline information for assessing and monitoring mast production in the park, especially its influence on black bear population dynamics. An abundant oak mast crop allows bears to gain sufficient weight for winter dormancy and cub production and results in fewer movements of bears out of the park in search of fall foods. A good mast crop also usually results in high reproduction of exotic wild hogs.

Have you ever noticed that most acorns picked up from the ground have small holes in them? When they are broken open, it is apparent that they have been eaten from within. Some still harbor a small white grub. Acorns are often attacked on the tree. The female acorn weevil (*Curculio* sp.) bores a small hole into the acorn with her elongated snout. She then lays an egg in the hole. The egg hatches into a small grublike larva that feeds on the acorn embryo. When a weevil larva matures, it bores its way out of the acorn and falls to the ground. The acorn moth lays her eggs in the weevil's exit holes. Her eggs hatch into larvae that feed on what the weevils have left of the acorn. Squirrels, chipmunks, and turkeys consume most of the sound acorns that fall to the forest floor. Acorns attacked by weevils and moths are discarded. Discarded acorns harbor a succession of tiny inhabitants before they decay completely into the forest floor.

Sassafras is a common tree at low and middle elevations. It is found in open oak and pine forests as well as in old fields and disturbed areas. Sassafras is unique in that it has three different types of leaves: unlobed, two-lobed, and three-lobed. These leaves, which are randomly arranged on the branches, are often referred to as a sock (unlobed), a mitten (two-lobed), and a glove (three-lobed). Yellowish-green flower clusters open in early April. Sassafras fruit (dark blue "berries" borne on bright red stalks) is attractive and distinctive. The fruits are highly prized by park bears,

One of the evergreen ferns in the park, the Christmas fern is easily identified by the lobes at the bases of the leaflets. Roots of this fern were used in preparations by Native Americans to treat toothaches, chills, digestive problems, rheumatism, and pneumonia.

who break many sassafras branches in their efforts to reach them.

Christmas ferns are common in the rocky pine and oak woodlands as well as in cove hardwood forests. The leaflets, which are "eared" at their bases, resemble mittens or Christmas stockings. Besides being an evergreen fern, the "eared" leaflets are a good field characteristic for identifying this species. Spores are borne on the undersides of small fertile leaflets along the upper portion of some of the fronds. The sterile leaflets below are distinctly larger. Newly emerged ferns (fiddleheads) rapidly unwind and unfold in the warming spring sunlight.

Newly emerged ferns are coiled and resemble the head of a fiddle, hence the name "fiddleheads." In the warmth of spring, they quickly expand into a more fern-like form.

Hemlock Forest

Hemlocks dominate streamsides and moist shady slopes throughout the Appalachians. Water temperatures remain cold year-round, which cools and dampens the air along streams. Hemlocks survive better in these conditions than any other species. Along streams and lower slopes up to an elevation of 3,500 to 4,000 feet, the eastern hemlock is a common tree. It also occurs on exposed slopes and ridges at middle elevations and up to almost 5,500 feet. Very few hemlocks occur above 5,500 feet. The Smokies has more eastern hemlock trees over 160 feet tall than anywhere else in the world. Several of the trees are six feet wide at the base.

The hemlock woolly adelgid (*Adeleges tsugae*) is moving south and west and is threatening every hemlock in the eastern United States. This insect was first discovered in the Smokies in 2002 and has been the focus of control efforts ever since (see chapter 11).

Associated with the hemlock are such trees as American beech, American holly, sweet and yellow birches, black and pin cherries, silverbell, cinnamon clethra, red and sugar maples, and the tuliptree (yellow-poplar). Rosebay rhododendron is an abundant streamside understory shrub, while Catawba rhododendron occurs on the higher exposed ridges. Other shrubs include dog-hobble, hobblebush, mountain laurel, scarlet elder, wild hydrangea, and thornless blackberry. The variety of spring-blooming herbs is much less than in the cove hardwood forests.

Only a few plants can tolerate the heavy shade of hemlock groves. Partridgeberry is one of them. The flowers grow in pairs with fused bases. As they mature,

Partridgeberry is a slender, evergreen creeper, which takes root along its prostrate stem. The small, shiny, oval leaves grow in opposite pairs. The flowers are also paired and give rise to the aromatic, red partridgeberry, or twinberry.

each pair of flowers produces a single red fruit. Another is the downy rattlesnake plantain. This plant is actually an orchid, but you have to look closely at the individual flowers to see that they do indeed resemble the more familiar orchids. The plant was probably named because of the resemblance of its leaves to a snake skin. Indians and mountain folk believed that the plant was effective in the treatment of snakebite. It is curious how traditional uses of plants are so often suggested by characteristics in their physical appearances.

The common polypody fern is a small evergreen fern with leathery leaves whose blades are lobed and connected at their bases (deeply pinnatafid). It thrives in the shade of the hemlock forest, usually growing on boulders, cliff faces, fallen logs, or other such places.

The Great Smoky Mountains National Park contains one of the most extensive old-growth forests in the eastern United States. Will Blozan of the Eastern Native Tree Society wrote the following:

> In spite of its relatively small size, the park, with nearly ideal growing conditions, high tree diversity, and protected ancient forests, likely has the highest concentration of record trees anywhere in the continental U.S. The vast regions of undisturbed forest and the retention of continuous natural processes has allowed for the development of ancient forests with very old and large trees. All the national champion trees from the Smokies are located in ancient or minimally disturbed forests and currently provide one of the best approximations of the quality and

Table 5.1. Twenty tallest and biggest (largest girth) species of trees in Great Smoky Mountains National Park.

Species	Tallest (Location)	Largest Girth (Location)
Eastern White Pine (*Pinus strobus*)	187.6 ft. Boogerman Loop Trail, NC	14 ft. 10 in. Half Acre Ridge, NC
Tuliptree (*Liriodendron tulipifera*)	178.2 ft. Baxter Creek Trail, NC	25 ft. 4 in. Middle Prong, TN
Eastern Hemlock (*Tsuga canadensis*)	173.1 ft. Big Fork Ridge, NC	19 ft. 1 in. Caldwell Fork, NC
Black Locust (*Robinia pseudoacacia*)	171.8 ft. Bradley Fork, NC	13 ft. 2 in. Walker Camp Prong, TN
White Ash (*Fraxinus americana*)	167.1 ft. Big Branch, Big Creek, NC	17 ft. 6 in. Indian Camp Creek, TN
American Sycamore (*Platanus occidentalis*)	162.3 ft. (tied) Big Creek, NC Lynn Camp Prong, TN	18 ft. 9 in. Buck Fork, TN
Pignut Hickory (*Carya glabra*)	157.5 ft. Big Fork Ridge, NC	12 ft. 2 in. Roaring Fork, TN
Yellow Buckeye (*Aesculus octandra*)	157.3 ft. Webb Creek, TN	19 ft. 1 in. (dead) Gabes Creek, TN
Bitternut Hickory (*Carya cordiformis*)	156.3 ft. Mouse Creek, NC	13 ft. 1 in. Porters Creek, TN
Red Spruce (*Picea rubens*)	155.3 ft. Styx Branch, TN	14 ft. 1 in. Breakneck Ridge, NC
Cucumbertree (*Magnolia acuminata*)	151.9 ft. Baxter Creek, NC	18 ft. 6 in. Messer Fork, NC
Northern Red Oak (*Quercus rubra*)	151.4 ft. Mouse Creek, NC	21 ft. 7 in. Double Gap Branch, NC
White Basswood (*Tilia heterophylla*)	150.3 ft. Baxter Creek, NC	14 ft. 0 in. Cataloochee Creek, NC
Shagbark Hickory (*Carya ovata*)	149.8 ft. Caldwell Fork, NC	11 ft. 3 in. Rich Mountain, TN
Biltmore Ash (*Fraxinus americana* var. *biltmoreana*)	148.6 ft. Mingus Creek, NC	12 ft. 2 in Cades Cove, TN
White Oak (*Quercus alba*)	147.1 ft. Winding Stairs Branch, NC	15 ft. 7 in. Cades Cove, TN
Sugar Maple (*Acer saccharum*)	144.2 ft. Baxter Creek, NC	14 ft. 4 in. Jones Branch, TN
Sweetgum (*Liquidambar styraciflua*)	142.8 ft. Big Creek, NC	11 ft. 4 in. Bullhead Branch, TN
Red Maple (*Acer rubrum*)	142.6 ft. Cannon Creek, TN	21 ft. 10 in. Indian Camp Creek, TN
American Beech (*Fagus grandifolia*)	142.6 ft. Mingus Creek, NC	13 ft. 8 in. Roaring Creek, TN

Data from Eastern Native Tree Society, August 2007.

size of trees that once existed in presettlement landscapes. In fact, several trees recently located in the Smokies now represent the maximum dimensions *ever recorded* for their species.

Blozan noted that exceptionally tall trees are the result of optimum growing conditions—rich soils, abundant rainfall, and moderate temperatures. It is extremely important to realize the value of a forest that in 2008 still sets new standards and shatters historical records. One white pine tree, locally known as the Boogerman Pine, stood 207 feet in 1995, the highest accurate measurement for any tree in the eastern United States within modern times. Unfortunately, Hurricane Opal in October 1995 caused the top to break and reduced its height to its present 187.2 feet. The Boogerman Pine stands alone as the only legitimate 200-foot tree in the East, and the white pine stands as the only eastern species living today that has been measured to heights of 180 feet or more.

The Eastern Native Tree Society in North Carolina maintains an up-to-date listing of accurate tree attributes for the eastern United States (Table 5.1). Two of these attributes are height and largest girth. The extreme accuracy of techniques used by members of the ENTS has revealed that the tallest trees in the park are generally those in second-growth forest. With the exception of eastern hemlock and red spruce, most of the trees native to the park will reach their maximum height before 100 years of age. Thus, due to logging in the early 1900s, trees in the Great Smoky Mountains National Park are likely now entering the period when they will be reaching their maximum height potential. As expected, the biggest (largest girth) trees are the older ones in the old-growth forests. These trees have had many years to bulk up and add wood. They have grown in a good location and have been able to maintain a decent growth rate for almost a century.

The *2006–2007 National Register of Big Trees*, a biennial American Forests publication, contains 16 national champions of 15 species from the Smokies (two are park co-champions). A national champion tree is a specimen that has the most points of any known and nominated tree of its species (Table 5.2). The point scale is based on three dimensions: circumference, height, and average crown spread. A tree can be listed as a co-champion if it is within five total points of another tree of the same species.

Will Blozan stated: "Some of the park's really big trees, such as the cucumber magnolia and red oak, *should* be national champions but the current tree on the *American Forests National Register of Big Trees* is mismeasured (overstated). As far as state record trees, GRSM probably contains over 200 legitimate champions, but as stated above the current record trees are overstated due to 1) mismeasurement and 2) multiple stems."

In December 2006 Blozan and his partner Jess Riddle found a grove of giant hemlocks—six trees, each more than 160 feet tall with some trunks measuring

Table 5.2. 2006–2007 National champion and co-champion trees in Great Smoky Mountains National Park with year of last measurement.

Species	Girth	Height	Spread	General Location
Allegheny serviceberry—1997 *Amelanchier laevis*	77"	101'	36'	Boulevard Prong Sevier Co., TN
Black cherry—1997* *Prunus serotina* var. *serotina*	169"	132'	51'	GSMNP, TN
Carolina silverbell—2001 *Halesia carolina*	152"	110'	43'	South of Grotto Falls Trail Sevier Co., TN
Catawba rhododendron *Rhododendron catawbiense*	15"	25'	15'	Baxter Crk. Tr., NC
Cinnamon clethra—1995 *Clethra acuminata*	10"	33'	12'	Caldwell Fork Trail
Devil's walkingstick—1997 *Aralia spinosa*	23"	74'	16'	Boulevard Prong Sevier Co., TN (dead) Haywood Co., NC
Eastern hemlock—1995 *Tsuga canadensis*	202"	165'	39'	Ramsay Branch Sevier Co., TN (dead)
Fraser magnolia—1998 *Magnolia fraseri*	118"	121'	33'	GSMNP, TN (dead)
Mountain laurel—2003 *Kalmia latifolia*	48"	25'	18'	Rabbit Creek Blount Co., TN
Pin cherry—1999 *Prunus pensylvanica*	58"	75'	41'	Brushy Mtn. Trail Sevier Co., TN
Red hickory—2001** *Acer rubrum*	276"	141'	88'	Maddron Bald Trail Cocke Co., TN
Red spruce—1997* *Picea rubens*	144"	146'	34'	Thicket Branch Swain Co., NC
Red spruce—1986* *Picea rubens*	169"	123'	39'	Breakneck Ridge Swain Co., NC (dead)
Striped maple—1997 *Acer pensylvanicum*	44"	77'	31'	Trillium Gap Trail Sevier Co., TN
Yellow buckeye—1995 *Aesculus octandra*	229"	136'	53'	Gabes Mtn. Trail Cocke Co., TN (dead)

From "National Register of Big Trees—2006–2007" published by American Forests.
*Co-champion.
**Possibly misidentified. Will Blozan of the Eastern Native Tree Society says this is a Bitternut hickory.

over 20 feet in circumference—along Big Fork Ridge in Cataloochee. Ground measurements indicated one tree was more than 170 feet tall, a height that Blozan calls "the Holy Grail for eastern hemlocks." This tree, named the Yuhgi Hemlock, was found to be 171.7 feet tall. In more than a decade of searching, he had found only 29 eastern hemlocks more than 160 feet tall—and never a 170-footer. The sad part is that none of the six newly discovered trees were alive, having fallen victim to the hemlock wooly adelgid (see chapter 11). As of August 2007, Blozan and his colleagues have documented 67 hemlock trees that are 160 feet high or taller in Georgia (1), South Carolina (5), Tennessee (4), and North Carolina (57). Twenty-eight of all the known trees over 160 feet are on the Big Fork Ridge.

In early 2004 Jess Riddle discovered within the park what was then the largest known eastern hemlock in the world. Climbed and measured by a direct tape drop in 2007, the tree, named the Laurel Branch Leviathan, was found to be 156.3 feet tall. It was also found to have 1,583 cubic feet of wood, more than any other known eastern hemlock. At 50 feet high, the girth measured 13.4 feet; at 25 feet high, it was 13.6 feet around; and at a little less than 5 feet above the ground, the trunk was 18.3 feet in circumference. Blozan stated: "This tree has very little taper. It maintains its thick trunk."

As of August 2007, Blozan has discovered four eastern hemlock trees in the park exceeding 170 feet. The tallest, known as the Usis Hemlock, is 173.1 feet in height, has a girth of 15.75 feet, and a volume of 1,533 cubic feet. In terms of volume, the Usis Hemlock is the fourth largest hemlock in the park. The tree with the largest girth (19.1 feet) and largest volume (1,601 cubic feet; greater than 80,000 pounds) is the Caldwell Giant. It stands 152.1 feet tall.

Effects of Fire

Forest fires are most frequent in the spring—after the winter rains have ceased and before the vegetation has advanced far enough to keep the ground cover from drying out—and in the autumn, after the leaves have fallen. Fires are rare during the season of active growth and cease entirely with the fall of snow and the scarcity of lightning strikes.

Since about 1940, the park service has suppressed fire, prompting concerns over possible changes in forest composition and structure. A study documenting changes in xeric forests in the park (that is, forests with little moisture) between 1936 and 1995 showed significant differences. Between the 1970s and 1995, canopy density on fire-suppressed and low-intensity fire sites remained relatively stable, while that on sites of high-intensity fires increased rapidly. During this period, abundant regeneration of pines occurred on some burned sites. On fire-suppressed sites, densities of shade-tolerant, late-successional species such as red maple, blackgum, eastern white pine, and eastern hemlock have increased, while the abundance of pioneer species such as Virginia pine and dogwood experienced significant declines. Changes in the canopies of xeric forests since the onset of fire suppression may alter response to future fire events and complicate the restoration of historical composition and structure in these communities.

The park allows naturally ignited fires to burn if they meet certain criteria and pose no threat to human safety, structures, or park-managed natural and cultural resources. Fire in areas of oaks and pines actually stimulates reproduction and is considered by ecologists to be a benefit in maintaining a naturally occurring ecosystem whose evolution was shaped by periodic fire. The relationship between fire and pine-oak forests is a self-maintaining cycle. Pines, being more drought-resistant than most deciduous trees, tend to grow on rocky ridges and south- and west-facing slopes, where the sun-heating effect is greatest. Along with some of the dry-slope understory plants, such as scarlet oak and laurel, pines burn readily. Thus, fires starting on south or west slopes, having good fuel, are more apt to spread than those starting on wetter slopes. At the same time, thick bark or resistant root systems enable these dry-slope plants to survive. The pine-oak forests simultaneously encourage fire (which operates to exclude other trees) and successfully resist destruction by it. Thus, they ensure their own continuance.

Table-mountain pine is a fire-dependent tree. In the southern Appalachians it occurs from low elevations up to about 5,000 feet, where it often is found in nearly pure stands. It apparently requires relatively high-intensity fire to reproduce, as the cones remain closed and "sealed" with the pine's resin for several years after they are produced. Should a fire of suitable intensity occur, the resin melts off and the cones open with an audible "pop," reseeding the burned area. While this characteristic is known in other pine species in North America, table-mountain pine is the only species to do so in the Appalachians. Although the cones will eventually open after several years, the seeds stand little chance of surviving in the dense leaf litter on the unburned ground. Only ground surfaces with little or no organic layer of leaf litter or decayed leaves ("duff") result in high rates of germination of the pine's seeds—like after a fire. The released seeds are usually flat (non-glossy) and black, perhaps mimicking the charcoal on burned ground. Many ground-feeding birds are attracted to fire sites, even while the flames are still burning. The seed's color and texture could have evolved due to birds eating seeds that were more visible.

The use of prescribed burning is a valuable technique for preserving natural diversity and forest health. The central purpose of the park's use of fire in the interior regions of the Smokies is to replicate as nearly as possible the role that naturally occurring fires played

Table 5.3. Acreage of prescribed burns, 2000–2007. Superintendent's Annual Reports.

Year	No. of Fires	Acreage
2000	1	664
2001	6	1,185
2002	5	838
2003	9	1, 352
2004	5	448.5
2005	5	2,440
2006	0	—
2007	4	3,000

From Superintendent's Annual Reports.

in shaping and maintaining the park's biologically diverse ecosystem. Controlled fires are employed in a variety of areas ranging from grasslands to forests for perpetuating rare plants; reducing hazardous fuels; ecosystem maintenance, including the control of woody species in fields; and encouraging native plant species, including the restoration of pine communities (Table 5.3). In 2000, for example, a total of 664 acres of grassland was burned in Cades Cove in an effort to reduce exotic fescues and restore native grasses. The park's second-largest-ever prescribed burn was a 1,034-acre pine restoration burn in March 2003. Known as the Arbutus Ridge Burn and located near Cades Cove, the burn area was historically a forest community with pine dominant in the overstory, a condition maintained in presettlement days by periodic fire. Changes in land management had become evident through changes in the stand structure, which had become a hardwood-dominant, closed-canopy forest. These changes had trickled down to all levels of the forest system. The Arbutus Ridge Burn significantly reduced some shade-tolerant plants that directly competed with the pines for space and sunlight. The largest prescribed burn in the park was the 2,300-acre Hatcher Mountain Burn in 2005.

Balds

Although there is no true timberline in the southern Appalachians, there are treeless areas on some of the higher mountaintops and ridges that are called balds. Even though trees are absent, other plants form a dense carpet over the balds. If the plants are largely shrubs belonging to the heath family, such balds are known as "heath balds"; if these plants are grasses and sedges, the balds are termed "grass balds."

Heath balds develop primarily on the windward side of the upper slopes and peaks above 3,000 feet after disturbance of the original forest. They typically develop in the northeastern part of the park, mainly within the limits of the spruce-fir zone, and have an evergreen canopy, deep leaf litter, and very acidic soil. The 421 heath balds in the park average 4.5 acres (1.8 ha) in size. They occur at mid- to higher elevations (94 percent occur at elevations between 3,600 and 5,200 ft), usually on extremely steep, knifelike rock ridges. Local inhabitants refer to heath balds as "laurel slicks," "woolly heads," "lettuce beds," "yaller patches," and "hells." From a distance, these areas may appear smooth or slick, but, in reality, they consist of tangles of mountain laurel and rhododendron, often 8 to 12 feet high and so thick that it is almost a solid mass of branches. They are nearly impenetrable except where the trail cuts through.

I remember my first experience in a heath bald. It was early in the summer of 1963, and we were hiking across Maddron Bald between Cosby and Greenbrier Pinnacle. This trail is not a high-priority trail, and the park maintenance crew from Cosby had not yet had a chance to clear it. The vegetation was about chest-high and completely obscured the trail. You could not see where you were placing your feet. We were relieved when we reached the far edge of the bald and fortunate that we had not encountered any obstacles, living or otherwise, while crossing it.

Years ago a mountaineer named Irving Huggins was trapped in a heath bald. It took him several days to find his way out. Since then, that area, an extremely rugged gorge that extends from Alum Cave Creek up to Myrtle Point on Mount LeConte, has been known as Huggins Hell.

A visitor once asked a mountaineer, "What would you do if you met a bear in one of those places?"

"Well," replied the mountaineer with a twinkle in his eye, "if I couldn't turn around and the bear couldn't turn around, there would be only one thing to do!"

"Yes?" questioned the visitor.

"When the bear opened his mouth, I'd stick my arm down his throat, grab him by the tail, and jerk him inside out! Then he'd be heading the other way!"

Catawba rhododendron, mountain laurel, Carolina rhododendron, rosebay rhododendron, and sandmyrtle are the dominant shrubs in heath balds. Since all are evergreen, these slicks remain green throughout the year. Usually around mid-June, these plants bloom in such profusion that they mask the green foliage. From a distance, a mountainside covered with these bushes appears to be a solid blanket of rose-colored bloom, and the sky seems to glow with its brilliance. This spectacle

Mountain laurel often forms dense thickets and is found at elevations up to 5,000 feet. It may sometimes grow to the size of a small tree. The green parts contain the poison andromedatoxin, which may prove fatal to sheep and other animals that eat the leaves. One helpful way to distinguish the large-leafed rhododendron from the smaller-leafed mountain laurel when not in bloom is to remember the phrase "long leaf, long name; short leaf, short name." Rhododendron was known to the early settlers as "laurel" and mountain laurel as "ivy."

Sandmyrtle, one of the dominant evergreen shrubs in heath balds.

can be seen from many vantage points along park roads, but is best experienced by hiking along portions of the Appalachian Trail or to Gregory Bald, Alum Cave Bluffs, or Mount LeConte.

Upon close examination, the mountain laurel flower consists of arched stamens held under tension by small pockets in the petals. When an insect visits a mature flower, the slightest nudge of a stamen frees its pollen-bearing head from its pocket. The stamen springs upright, dusting the insect visitor in a shower of pollen.

Reindeer lichen (*Cladina* sp.) is usually found on high rocky outcrops, in heath balds, and in dry open woodlands. It is normally dry and brittle, but becomes soft and spongy in wet weather.

The origin of heath balds has been a matter of conjecture and discussion for many years. Scientists have speculated about what created them, how old they are, and what maintains them in the face of otherwise rapid forest succession in the rest of the park. Heath balds are considered stable communities with comparisons of selected heath balds on 1930s' and 1980s' aerial photographs showing no changes in area. Three possible theories as to their origin have involved fire, windfall, and landslide. More recently, bald occurrence has been positively correlated with burned sites, old-growth conditions, and a highly acidic rock type. Long before man's appearance in the Smokies, lightning strikes resulted in devastating fires that cleared the areas. In addition, intense fires consuming logging slash often resulted in severe soil erosion. Mountain laurel and rhododendron shrubs were able to resprout in the burned-over areas more easily than the trees and thus became the dominant plants before the trees could reestablish themselves. Once established, this shrub community, with its dense evergreen canopy and thick, slowly decomposing acidic leaf litter, is resistant to tree invasion.

Beginning in 2004, Dr. Rob Young of Western Carolina University dug pits to bedrock in each of 14 heath balds and found them to be essentially dry peat lands. This circumstance is in itself very unusual, since peat lands are usually found in low wet depressions. The soils are highly organic, extremely acidic (with pH below 3), and have a very low base saturation. Aluminum saturation is very high, giving the soils a low productivity rating. Radiocarbon dates showed the oldest ones to be nearly 3,000 years old, although others appear to be much younger. Evidence of charcoal was found in the lowest layers of most balds, and, in at least one instance, a layer of charcoal was found part of the way down the soil profile, indicating that a heath bald existed for a long time, then burned completely, but returned as a heath bald, possibly short-circuiting the normal process of succession.

While heath balds are practically impenetrable, grass balds are mountaintop meadowlands comprising grasses, sedges, and various other herbs. They are the most limited vegetation type of the Smokies and one of the most distinctive. Grass balds are found mostly in the southwestern part of the park on rounded tops or slopes between 4,500 and 5,700 feet. They are surrounded by deciduous forest. Grass balds are from 100 to 300 yards wide and from several hundred yards to two miles in length. Gregory, Parsons, Silers, and Andrews balds are the best-known examples. The origin of the grass balds has never been satisfactorily explained. The Indians may have cleared these originally, perhaps as gathering places for religious ceremonies or so they might better watch the movements of an enemy tribe. Early settlers may have cleared the balds for grazing, and cattle, sheep, and horses kept them clear of trees. Or landslides and violent wind patterns may have erased the ancient tree cover.

Brewer (1993) noted:

> Keith Langdon of the National Park Service a few years ago obtained a copy of the old field notes of William Davenport, who in 1821 surveyed the line between Tennessee and North Carolina, from Davenport Gap, near the eastern end of the Smokies, south to the Georgia line. He went right down the crest, across the site of every bald except Andrews.
>
> Nowhere in his notes did Davenport hint at a bald until he reached what came to be called Gregory Bald. Then he mentioned "the top of the bald in sight of Tellessee Old Town" (the former Cherokee town of Tallassee beside the Little Tennessee). After another mile, they came to a "Red oak . . . in the edge of the second bald spot." This would have been Parson Bald.
>
> Davenport's notes should be reasonable proof that settlers did not create Gregory and Parson balds—but if one is to use them for that purpose, one also must accept them as evidence that the other state-line balds in the Great Smokies did not exist in 1821.

In 1883 William Zeigler remarked: "Every spring thousands of cattle, branded, and sometimes with bells, are turned out on these upland pastures." Q. R. Bass (1977) stated:

> The most dramatic evidence for the influence of grazing as a causal factor in bald formation is to be found on Hemphill Bald on the southeastern boundary of the park. The park boundary fence divides this grassy bald approximately into two halves. On the private side of the bald, where horses and cattle are still allowed to roam freely, the grassy bald vegetation and the open "orchard" cover still exist. Conversely, the park side of the bald is completely and densely overgrown in pine and oak forest. One is therefore forced to conclude from this evidence that the present balds are the result of, or at least were perpetuated by, the grazing of European-introduced domesticated animals. It is also evident that the summit zone flora is sensitive to any stress imposed upon it. The exact influences that prehistoric wildlife (especially deer and elk) and aboriginal man exerted on the summit zone vegetation are unknown.

A study of the grass balds in 1931 led Stanley Cain to conclude that they were natural phenomena for the "soil profiles show from a few inches to a foot or more of homogenous [sic] black soil of grassland type, which is too deep and mature to have developed since the advent of the white man, with the possibility of his having cleared off the trees." In another 1931 paper, W. H. Camp theorized that Gregory Bald was not originally a pure grassy meadow, but one with numerous "shrub islands" of various types. This view was substantiated by Mrs. John Oliver, a resi-

Cattle, sheep, and horses formerly grazed the grass balds. 1930.

Flame azalea, which is also known as wild honeysuckle to many mountain people, occurs as single plants or in scattered clusters throughout the park. It blooms from April to July, depending on elevation.

Flame azalea (close-up).

dent of Cades Cove, who said that she "had it from the older folks long dead" that Gregory Bald was "originally a blueberry meadow" and had "always been a bald." Thus, Camp concluded that the grassy balds were a natural phenomenon probably produced by occasional desiccating southwesterly winds, their grassland character being intensified during the last century by fires and overgrazing. Once the Smokies began to be settled, clearing operations together with cattle- and sheep-grazing probably served to keep these meadows in an open condition. Cattle last grazed on Gregory Bald in 1936. Since that time, plants have been invading the bald from the surrounding forest. Along the edges of some grass balds, such as Gregory Bald, large concentrations of flame azalea exist. When in bloom during the last half of June (usually between June 20 and 25), they are a spectacular sight.

Mountain oat grass is the characteristic grass of the park's high balds. Its name comes from its resemblance to oats. When its flowers ripen, they fall to the ground. Their bristles—sensitive to the amount of available moisture—twist and turn as moisture changes. This movement drives the flower and its contained seed through the litter and into the soil—a self-planting seed.

In 1983 the National Park Service made the decision to intervene with natural succession in order to maintain two grass balds—Gregory and Andrews. The two balds are being managed with the "primary objective being the preservation of their distinctive plant composition and scenic value." Maintenance was begun on Andrews Bald in the summer of 1983 and on Gregory Bald in the summer of 1984. Each year, encroaching vegetation is cut by Park Resource Management vegetation crews to preserve the native grass, forb, and azalea plant communities. Blackberries, hawthorn, and woody saplings are controlled using weed-eaters and a front sickle bar mower.

In 1999, a former student and I backpacked 50 mammal live-traps from the Clingmans Dome parking area (Forney Ridge) to Andrews Bald, a distance of 1.8 miles. To my knowledge, this was the first-ever effort to sample small mammal populations on Andrews Bald. On the hill slope along one margin of the bald is a small "hanging bog" where water seeps out, spreads out, and is retained on the slope. Mosses and other wetland plants grow here. We set the traps while listening to the howling of coyotes in the valley below, and then we hiked out. We repeated the procedure for the next two days in order to check and rebait the traps and

In June and July, Andrews Bald is beautiful with blooming flame azaleas and Catawba rhododendrons. It is also the site of a small bog.

Spence Field, a grass bald along the Appalachian Trail in the western portion of the park.

then to pick them up and backpack them out. Thus, we made three roundtrips to Andrews Bald in three days. In addition to the beautiful scenery and views from the bald, we discovered deer mice, white-footed mice, southern red-backed voles, and three species of shrews. The identification of one shrew was later confirmed by a taxonomist at the National Museum of Natural History in Washington, D.C. as a northern water shrew (*Sorex palustris*). At 5,800-feet elevation, this represents the highest elevation known for this species within the park. It also represented, at that time, only the second specimen from the North Carolina portion of the park. The first was taken by me along Beech Flats Prong (4,000 feet) near its intersection with U.S. Route 441 (transmountain road) in Swain County in 1980.

CHAPTER 6

ANIMALS

We remain important, you and I and all mankind. But so is the butterfly—not because it is good for food or good for making medicine or bad because it eats our orange trees. It is important in itself, as a part of the economy of nature.

—Marston Bates (1960)

The park is home to approximately 6,000 known species of nonmicrobial invertebrates and to 79 native fish species, 43 species of amphibians, 40 varieties of reptiles, more than 240 species of birds, and 66 types of mammals. It is the center of diversity for lungless salamanders (Family Plethodontidae).

Invertebrates (Terrestrial)

Millipedes (Class Diplopoda)

Millipedes are common throughout the park. They feed on decaying vegetation and other soft plant materials, and they play a major role in decomposing the estimated 500,000 tons of leaves that fall to the ground in the park each year. Millipedes differ from centipedes in that they possess two pairs of legs on each body segment; centipedes have only one pair of legs per segment. A millipede embryo, however, has only one pair of legs to every body segment. In the adult, the first four (thoracic) segments remain single, but the other (abdominal) segments fuse in pairs, so that each adult ring represents two embryonic segments

> **Do You Know:**
> Where monarch butterflies overwinter?
> Which species of trout is native to the park?
> Which is the largest species of salamander in the park?
> How many types of venomous snakes reside in the park and what they are?
> What is the largest species of woodpecker in the park?
> How many species of mammals occur in the park?

The centipede's body is divided into two regions—a head and a trunk. Centipedes are carnivorous, with the legs of the first trunk segment being modified into large poison claws.

and has two pairs of legs; hence the technical name for this group, Diplopoda. Most millipedes have scent glands. These discharge obnoxious substances that discourage would-be predators. Hydrocyanic acid, iodine, and quinine are the most common active ingredients of these emissions. You can easily detect the bitter-almonds smell of hydrocyanic acid by confining an agitated millipede in the cup of your hands or in a small container.

Snails (Class Gastropoda)

Snails, another major decomposer of plant and animal matter, are recognized for their role in cycling calcium through natural systems. About 150 species have been documented in the park. Snails bear two pairs of tentacles on their head end. Simple eyes are located at the tips of the larger pair. Snails thrive on wet conditions, humid air, and a rich supply of plant food. Therefore, they are most active at night and in

Millipedes feed largely on decaying organic matter and play a valuable role in decomposing plant material. They protect themselves by curling up so that the hard parts of their back cover their leg-bearing undersurface.

Millipedes are common at all elevations. Eggs are laid in clusters in damp earth. When young first hatch, they have only three pairs of legs; as they go through successive molts, the number increases until most species have 30 or more pairs.

Land snails, found among forest debris, feed upon plant material rasped free by their tongue. They can detect food by scent.

Walkingsticks are large, usually wingless insects with legs all about the same length. They live and feed on trees such as oak, locust, cherry, and walnut. The female's 100 or so eggs are dropped singly to the ground to hatch the following spring.

damp weather. Snails move by the muscular contraction of their large "foot." Movement is made easier by a track of slime produced by a gland within the foot.

Insects (Class Insecta)

The four most abundant and speciose groups of insects in the park are the Lepidoptera (moths, butterflies, and skippers), Coleoptera (beetles), Diptera (flies), and Hymenoptera (wasps, ants, bees, etc.). As of January 2008 the total number of Lepidoptera species known from the park exceeds 1,300; the total number of beetle species is approximately 2,140; the total number of fly species documented to date is over 700; and the total number of hymenopteran species so far is 699.

One interesting member of the beetle order is the burying beetle (*Nicrophorus tomentosus*). When a small mammal such as a shrew or a mouse dies, burying beetles dig the soil out from beneath the carcass, piling the excavated dirt above it. Slowly, the carcass sinks into the soft earth. Within several hours, it completely disappears. With the carcass buried, the beetles lay their eggs within it. The larvae hatch and feed on the decaying flesh. Along with fungi, bacteria, millipedes, and snails, these beetles play a valuable role in the process of decomposition and recycling of nutrients in the ecosystem.

Many decaying logs in the damp woodlands are laced through with the burrows and galleries of the woodroach (*Crytocercus punctulatus*). These primitive insects live in small colonies within the decaying wood. The woodroach eats wood, but it cannot digest it. For this, it depends upon colonies of single-celled protozoans that live within its gut. Without the protozoans, the woodroach would starve. For their part, the protozoans appear to depend totally on the roach to gather the wood fibers they require; they have never been found outside of the woodroach's gut. Woodroaches increase in size until they reach maturity at about six years of age.

Synchronous firefly beetles (*Photinus carolinus*) are abundant and common in various watersheds throughout the park. Perhaps best seen near Elkmont, they have become a "celebrity insect," with an increasing number of visitors (up to 2,000 people per night) wishing to view the "waves" of light. The waves occur from thousands of the beetles flashing five to six times within the same several-second period, repeatedly, late at night. The greatest activity normally occurs during the month of June. Firefly larvae are predaceous and feed on various small insects and on snails.

The many milkweed plants growing in clusters along roadsides and the edges of meadows are named for their distasteful and poisonous milky juice. They are shunned by many mammals. Caterpillars of the monarch butterfly (*Danaus plexippus*), however, are immune to the milkweed's poisons. Because their bodies absorb large quantities of the noxious milkweed substances, they are distasteful to birds. The caterpillar's bright and bold markings advertise their distastefulness to would-be predators—offering it a degree of protection. Adult monarchs retain the noxious qualities of the caterpillar. Their conspicuous orange and black coloring is also a warning pattern that protects them.

Although a great deal seems to be known about the monarch butterfly, many unsolved mysteries remain about this species. In all the world, no butterflies migrate like the monarchs of North America; it is a migration that includes the countries of Canada, the United States, and Mexico. Some individuals travel up to 3,000 miles, farther than any tropical butterflies. Somehow, they return to the same winter roosts and often to the exact same trees, even though the same individuals do not return each fall. In March, as temperatures begin to rise, the monarchs become active again. After they have mated, the butterflies—mostly females—head north. By May or June, most of the monarchs' northern journey has come to an end. The

Tagging of monarch butterflies takes place in the park during September and October.

females lay their green eggs on milkweed plants and then die. About one week later, the eggs break open to reveal a yellow, black, and white striped caterpillar. The caterpillar will first eat its own eggshell, then eat the leaves of its food plant. It is the children's grandchildren of the butterfly that left Mexico in the spring that migrate south the following fall. Monarchs can live from three weeks to nine months.

Monarchs are usually sighted in the park for the first time each year in late March or early April. The earliest recorded sighting was on March 16, 1953, at park headquarters by Arthur Stupka. These adults mate and the females lay eggs. The offspring leave this area, but it is not known exactly where they go. There are few, if any, monarchs in the park from mid-June until early August when they reappear. These adults breed, and their offspring join in the fall migration to the oyamel fir forests of Mexico's Sierra Madre Mountains. The latest record of this species in the park is November 14, 1949, at park headquarters, also by Arthur Stupka. Some monarchs that leave eastern Canada and the eastern United States may cover a distance of 2,800 to 3,000 miles during the fall migration.

The park provides habitat that supports both monarch reproduction and migration. Several species of milkweed occur in the park that can serve as host plants for monarchs. Most significantly, there are several large patches in Cades Cove on which monarch eggs and larvae can be found, usually starting in May. As is the general pattern for the southern United States, few, if any, eggs and larvae can be found in mid-summer. However, one or two additional generations are produced in late summer and early fall. In the park, the highest densities of eggs and larvae seem to occur in these later generations, when researchers have sometimes seen more than one monarch per milkweed ramet (stem). Adult monarchs are most visible in the park from late September through mid-October as they migrate to their overwintering sites in Mexico. Fall flowers in Cades Cove fields, at Purchase Knob, and other locations within the park are also important monarch habitats, providing a refueling opportunity for monarchs on their journey.

Beginning in 1999 an adult monarch butterfly tagging program was begun in the park. Over the years it has been supervised by Nancy Keohane, Janice Pelton, Michelle Prysby, Wanda DeWaard, Paul Super, and Jason Love. Several hundred butterflies have been caught in the park, tagged, and released on their way to Mexico where they overwinter and hopefully are recaptured. In 2006, 121 monarchs were tagged in Cades Cove, and in 2007, a record 185 were tagged. Most of the captures in the park have come from Cades Cove, although some have been taken at Twin Creeks and Purchase Knob. They have been captured while feeding on plants such as thistle, Joe-Pye-weed, and red clover. All tagging data is sent to Monarch Watch, an international program that serves as a clearinghouse for information on monarch butterflies. Monarch Watch can be accessed on the Web at http://www.monarchwatch.org.

On September 21, 1998, a male monarch that was found as a caterpillar and reared by Wanda DeWaard was released on Caney Creek Road in Pigeon Forge. This road is directly next to Park property just off the spur between Gatlinburg and Pigeon Forge. It was recovered on March 10, 1999, in El Rosario, Mexico, 1,504 miles from its release point.

On September 28, 2001, 73 monarch butterflies were tagged in Cades Cove. One of those tagged by Janice Pelton, tag number ABT 239, was recovered by Bernardo Garcia in El Rosario on January 21, 2002, after covering a distance of at least 1,476 miles. This represents the first capture of a monarch butterfly tagged within the park.

Monarchs have an uncertain future due to loss of overwintering habitat in Mexico. An estimated 250 million may have frozen to death in the Sierra Chincua and El Rosario overwintering colonies following a severe winter storm on January 12–13, 2002. Up to 80 percent of the colonies may have died as a result of this severe weather combined with the diminished forest cover. The forest canopy was too thin to protect the delicate monarchs from the rain and cold weather. A healthy and intact forest serves both as an umbrella and a blanket that protects the monarch colonies from the wind, rain, and cold.

Monarch butterflies also face a new threat in the United States. Bt (*Bacillus thuringiensis*) corn is a genetically modified corn that contains a toxin to which lepidoptera (butterflies and moths) larvae are susceptible. Widespread planting of Bt corn began in 1997. Twenty-five million acres of Bt corn, 32 percent of the U.S. corn crop, were distributed throughout corn-growing regions by 1999. The pollen from Bt corn can prove fatal to monarchs if it covers milkweed on which monarch larvae are feeding.

On July 6, 2006, the U.S. Fish and Wildlife Service, the National Park Service, Canada's Wildlife Service and Parks Agency, and Mexico's Secretariat of the Environment and Natural Resources designated 13 wildlife preserves as protected areas. The Trilateral Monarch Butterfly Sister Protected Area Network will develop international projects to preserve and restore breeding, migration, and winter habitat.

Invertebrates (Aquatic)

Although numerous small wetland areas exist, there are few standing (*lentic*) bodies of freshwater such as ponds and swamps. Those that do exist, such as Gum Swamp and Gourley Pond (a temporary pond), both in Cades Cove, provide important habitat and breeding sites for many species of invertebrates as well as vertebrates such as amphibians and aquatic turtles. Most of the aquatic habitats in the park consist of flowing (*lotic*) systems such as streams and rivers. Over 2,000 miles of streams and rivers flow throughout Great Smoky Mountains National Park. In all but one case (Chilhowie Mountain), the source of water for every stream is located within the park; as the water flows away from its source and reaches lower elevations, it leaves the park.

Streams all arise directly or indirectly from precipitation deposited on the land as a part of the hydrologic cycle. They represent the excess precipitation over that held by the soil. Streams act as agents of transportation, carrying soluble materials in solution and fine grains of insoluble matter in suspension (mud), as well as rolling sand and gravel along the bottom.

Freshwater streams and rivers change over their course from being narrow, shallow, and relatively rapid to become increasingly broad, deep, and slow moving. Most streams are characterized by a repeating sequence of rapids and pools that decrease in frequency downstream. The headwaters of many park streams originate at high elevations in the spruce-fir forest. Their initial gradients are steep, waterfalls are abundant, and they lack tributaries. Such streams are known as first-order streams. When two first-order streams unite, the resulting waterway becomes a second-order stream; when two second-order streams unite, they form a third-order stream, and so on until they form a river. A mountain stream tumbling over stones in its path is usually cool and well oxygenated; as the water moves downstream and becomes more sluggish, the oxygen level tends to drop. Because of the continual addition of nutrients and detritus en route, nutrient levels tend to be higher downstream.

Sources of nutrients in headwater streams are limited to detritus such as leaves and woody debris, animal feces, dead insects, and dissolved organic matter from nearby vegetation that washes off during rain events. These streams generally lack suitable resources (food, cover) for fish or certain salamanders. As streams descend in elevation and become larger, the current slows somewhat, silt and decaying organic matter accumulate on the bottom, and algae and various aquatic plants are present. Such streams now contain logs, rocks, and boulders and provide habitat for many types of aquatic organisms including invertebrates, trout, and salamanders. Whatever the organisms living in or along a stream add to the water is continually exported downstream. As streams continue to descend in elevation, they become suitable for an even wider variety of aquatic plants as well as invertebrates, many species of fishes, salamanders such as the hellbender, turtles, and mammals such as river otters and beaver.

Bacteria and fungi serve as primary decomposers of organic material in stream systems. Other major feeding groups include the shredders, collectors, scrapers, and predators. Shredders such as cranefly larvae, caddisfly larvae, and crayfish feed on leaves and other large organic particles. Collectors include blackfly larvae and midge larvae, which filter fine detrital particles that settle on the stream bottom. Scrapers such as other types of caddisfly larvae and water pennies (the larvae of a particular type of beetle) feed on the algae coating the rocks. Shredders, collectors, and scrapers are fed upon by predaceous insects such as stonefly and dobsonfly larvae as well as fishes such as darters, sculpin and trout, and aquatic salamanders. In order to keep from being swept away by the current, animals must either attach themselves to solid surfaces, swim strongly, or avoid the current by going under rocks or even down into the rubble on the bottom. Many of the immature insects (larval forms) are streamlined to reduce drag and have legs adapted for clinging to

Becky Nichols, NPS entomologist, and Ted Grannan, former seasonal employee, catching aquatic insects. Photo by Steve Bohleber.

the substrate. These larvae will ultimately metamorphose and develop into mature terrestrial (winged) forms in completing their life cycles.

Aquatic Macroinvertebrates

Adult caddisflies (Trichoptera) are usually small, mothlike creatures, active by night and seldom conspicuous. The larvae of most caddisflies build tubelike cases in which they live. These are made from sand, pebbles, shells, plant fragments, or other stream debris which the larvae glue together with silk they produce from their mouths. Many of these cases are portable, and the larvae drag them along as they move over the stream bottom. Others are fastened with silk to the stream bed, their open ends directed into the current. Many caddisfly larvae spin silken webs that snare food particles from the passing current.

Beginning in 1993 the park established 27 long-term aquatic macroinvertebrate sampling sites. These samples are providing long-term data regarding the status of the aquatic community, which in turn provides information on the health of the aquatic ecosystem. In 2002, a biological inventory of Gregory Cave was completed. The cave bioinventory resulted in finding 46 cave-dwelling arthropods and crustaceans, including 26 that had not been discovered in previous, sporadic surveys. In addition, two species of invertebrates were new and undescribed to science.

More in-depth information about specific invertebrates found in the park can be found at http://www.discoverlifeinamerica.org/atbi/species/animals/invertebrates.

Vertebrates

Fish

A total of 87 fish species in 15 families have been identified within the park. This includes 80 native species and seven introduced species. The National Park Service considers 10 of the 87 species to be extirpated.

Checklist of Park Species
- Family Atherinopsidae—Silversides (1)
  - Brook silverside (*Labidesthes sicculus*)
- Family Catostomidae—Suckers (7)
  - White sucker (*Catostomus commersoni*)
  - Northern hogsucker (*Hypentelium nigricans*)
  - River redhorse (*Moxostoma carinatum*)
  - Black redhorse (*Moxostoma duquesnei*)
  - Golden redhorse (*Moxostoma erythrurum*)
  - Shorthead redhorse (*Moxostoma breviceps*)
  - Sicklefin redhorse (*Moxostoma sp.*)
- Family Centrarchidae—Sunfish, Bream, Sun perch (11)
  - Rock bass (*Amploplites rupestris*)
  - Redbreast sunfish (*Lepomis auritus*) [introduced]
  - Green sunfish (*Lepomis cyanellus*)
  - Warmouth (*Lepomis gulosus*)
  - Bluegill (*Lepomis macrochirus*)
  - Longear sunfish (*Lepomis megalotis*) [extirpated]
  - Redear sunfish (*Lepomis microlophus*)
  - Smallmouth bass (*Micropterus dolomieu*)
  - Largemouth bass (*Micropterus salmoides*)
  - Spotted bass (*Micropterus punctulatus*)
  - Black crappie (*Pomoxis nigromaculatus*)
- Family Clupeidae—Herrings and Shad (1)
  - Gizzard shad (*Dorosoma cepedianum*)
- Family Cottidae—Sculpins (2)
  - Mottled sculpin (*Cottus bairdi*)
  - Banded sculpin (*Cottus carolinae*)
- Family Cyprinidae—Minnows, Shiners (28)
  - Central stoneroller (*Campostoma anomalum*)
  - Largescale stoneroller (*Campostoma oligolepis*)
  - Goldfish (*Carassius auratus*) [introduced]
  - Rosyside dace (*Clinostomus funduloides*)
  - Smoky dace (*Clinostomus funduloides* ssp.)
  - Whitetail shiner (*Cyprinella galactura*)

Spotfin shiner (*Cyprinella spiloptera*) [extirpated]
Common carp (*Cyprinus carpio*) [introduced]
Spotfin chub (*Erimonax monachus*)
Flame chub (*Hemitremia flammea*)
Bigeye chub (*Hybopsis amblops*)
Striped shiner (*Luxilus chrysocephalus*)
Warpaint shiner (*Luxilus coccogenus*)
River chub (*Nocomis micropogon*)
Emerald shiner (*Notropis atherinoides*) [extirpated]
Tennessee shiner (*Notropis leuciodus*)
Silver shiner (*Notropis photogenis*)
Rosyface shiner (*Notropis rubellus*) [extirpated]
Saffron shiner (*Notropis rubricroceus*)
Mirror shiner (*Notropis spectrunculus*)
Telescope shiner (*Notropis telescopus*)
Mimic shiner (*Notropis volucellus*) [extirpated]
Fatlips minnow (*Phenacobius crassilabrum*)
Stargazing minnow (*Phenacobius uranops*)
Tennessee dace (*Phoxinus tennesseensis*)
Longnose dace (*Rhinichthyes cataractae*)
Orangeside dace (*Rhinichthyes obtusus*)
Creek chub (*Semotilus atromaculatus*)

Family Fundulidae—Topminnows (1)
Northern studfish (*Fundulus catenatus*) [Occurs just downstream from the park boundary. Originally occurred in lower Abrams Creek, but has apparently been extirpated from that area.]

Family Ictaluridae—North American catfishes (8)
Black bullhead (*Ameiurus melas*)
Yellow bullhead (*Ameiurus natalis*)
Blue catfish (*Ictalurus furcatus*)
Channel catfish (*Ictalurus punctatus*)
Smoky madtom (*Noturus baileyi*)
Yellowfin madtom (*Noturus flavipinnis*)
Stonecat (*Noturus flavus*) [extirpated]
Flathead catfish (*Pylodictis olivaris*)

Family Lepisosteidae—Gars (1)
Longnose gar (*Lepisosteus osseus*)

Family Moronidae—Temperate basses (1)
White bass (*Morone chrysops*)

Family Percidae—Perch, Darters (19)
Greenside darter (*Etheostoma blennioides*)
Greenfin darter (*Etheostoma chlorobranchium*)

    Fantail darter (*Etheostoma flabellare*)
    Tuckasegee darter (*Etheostoma gutselli*)
    Blueside darter (*Etheostoma jessiae*) [extirpated]
    Duskytail darter (*Etheostoma percnurum*)
    Redline darter (*Etheostoma rufilineatum*)
    Tennessee snubnose darter (*Etheostoma simoterum*)
    Swannanoa darter (*Etheostoma swannanoa*)
    Wounded darter (*Etheostoma vulneratum*)
    Banded darter (*Etheostoma zonale*)
    Yellow perch (*Perca flavescens*) [introduced]
    Tangerine darter (*Percina aurantiaca*)
    Blotchside darter (*Percina burtoni*) [extirpated]
    Logperch (*Percina caprodes*)
    Gilt darter (*Percina evides*)
    Olive darter (*Percina squamata*)
    Sauger (*Stizostedion canadense*) [extirpated]
    Walleye (*Stizostedion vitreum*)
Family Petromyzontidae—Lampreys (2)
    Mountain brook lamprey (*Ichthyomyzon greeleyi*)
    American brook lamprey (*Lampetra appendix*)
Family Poeciliidae—Livebearers (1)
    Eastern mosquitofish (*Gambusia holbrooki*) [introduced]
Family Salmonidae—Trout (3)
    Brook trout (*Salvelinus fontinalis*)
    Rainbow trout (*Oncorhynchus mykiss*) [introduced]
    Brown trout (*Salmo trutta*) [introduced]
Family Sciaenidae—Drum (1)
    Freshwater drum (*Aplodinotus grunniens*)

There are approximately 2,115 miles of streams in the park, ranging in elevation from 874 feet to over 6,600 feet. Principal drainages are the Little River, Little Pigeon River, Pigeon River, and the Little Tennessee River, with Abrams Creek and the Oconaluftee River being tributaries of the Little Tennessee River. Impacts on park fishes may be caused by acid rain, accidental introduction of exotic species, temperature changes due to atmospheric warming as a result of greenhouse gases, or changes in the terrestrial environment that may impact aquatic ecosystems.

Annual or biannual fish population surveys are conducted in the larger streams in the park. A 100- to 200-meter portion of a stream is blocked off with nets and electroshocked in an attempt to remove all fish from that portion of the stream. Electroshocking simply stuns fish; it does not injure them. Fish are sorted by species, weighed and measured, and held in live boxes. This process is repeated two more times, after which the block nets are removed and the fish are released.

Electroshocking fish in a park stream.

Reliable estimates of populations of all common fish species can be calculated based on the reduced number of specimens of each species (depletion) taken during each pass (three-pass depletion estimates).

Trout

Brook trout once populated at least 550 miles of the Smokies' 750 miles of fishable streams. Before the park was established, factors such as logging and the resulting silting that clogged park streams, inadequately regulated fishing, and the introduction of the brown trout from Europe and the western rainbow trout brought about a reduction in numbers and distribution of the brilliantly colored native brook trout. Rainbow trout, introduced into nearly every watershed in the early 1900s, have established themselves as the dominant game species in Great Smoky Mountains National Park and have encroached upon many brook trout populations. Organisms occupying the same geographical region without interbreeding are said to be "sympatric." When brook trout distribution data from the year 2000 was compared to data collected by Willis King in 1936–37, it was possible to assess change over a 60-year period. Results vary by watershed; however, in 17 watersheds surveyed between 1992 and 2001, sympatric brook trout populations lost 0.5–3.5 km of range per stream to rainbow trout during the 60-year period. In most sympatric populations, it appears that rainbow trout do not systematically replace brook trout but rather have reached a point where the populations ebb and flow depending upon environmental conditions.

Between 1992 and 1995, 47 brook trout populations throughout the park were genetically typed using diagnostic allozyme loci and starch gel electrophoresis.

The native brook trout, known to mountain people as the speckled trout because of the conspicuous wormlike dark markings on its olive green back and dorsal fin, once occupied hundreds of miles of fishable streams in the park until its populations were diminished by siltation and rising temperatures caused by logging, excessive fishing pressure, and competition with the introduced rainbow trout.

Originally native to the western United States, the rainbow trout has been widely introduced throughout the world.

Of these populations, 64 percent (30) were pure southern Appalachian strain, 2 percent (1) were pure northern hatchery strain, and 34 percent (16) were hybrid populations. The presence of hybrid and northern populations closely follows historic stocking records of northern-strain brook trout in the park. Brook trout studies throughout the Appalachians from Tennessee north through Virginia indicate a distinct genetic break at the New River watershed in Virginia. Brook trout in New River and farther south constitute a distinct southern Appalachian genotype, while brook trout north of the New River watershed constitute a northern hatchery strain genotype. Based upon this work, the park has adequate information to substantiate the presence of a specific genotype present in the southern Appalachians and to protect the genetic integrity of these populations.

Long-term monitoring of brook trout throughout the park from 1998 to 2001 indicated that populations remained relatively stable despite four consecutive years of severe drought. During 2000 and 2001, the number of brook trout per mile of stream ranged from 240 to 3,552 fish/mile; however, most streams averaged 600 to 1,800 brook trout/mile. During this same period, brown trout densities ranged from 24 to 2,640 fish/mile, while rainbow trout densities ranged from 600 to 9,112 fish/mile throughout the park. Floods and droughts are by far the largest influence on brook trout populations in the park. Severe floods have reduced young-of-year production in some populations by 30 to 90 percent. Severe droughts reduce water levels, crowd adults, increase stress, and eventually increase adult mortality. Young-of-year brook trout appear to benefit from the lower water levels during droughts, whereas young-of-year rainbow trout appear to be negatively affected.

A brook trout restoration project was begun in 1998 on LeConte and Pilkey creeks. By 2000, monitoring surveys indicated excellent reproduction as young-of-year brook trout made up 56 percent (178/320) of the total brook trout catch in LeConte Creek and 67 percent (169/252) of the catch in Pilkey Creek. Self-sustaining brook trout populations in these two streams were well established by 2002–2003.

As of 2006, fisheries biologists have been able to restore 11 stream segments totaling 17.2 miles to pure brook trout populations. These include portions of LeConte Creek, Pilkey Creek, Winding Stair Branch, Mannis Branch, Ash Camp Branch, Sams Creek, Bear Creek, and Indian Flats Prong. Several more streams are being considered for restoration. Brook trout now coexist with non-native trout in another 69 miles of streams.

Adult brook trout in the park rarely live beyond four years of age and seldom exceed 200 mm (eight inches) in length. In 2000, brook trout monitoring surveys collected 4,232 brook trout, of which 111 (less than 3 percent) were greater than seven inches (178 mm). In 2001, monitoring surveys collected 2,095 brook trout, of which 74 (less than 4 percent) were greater than seven inches, and 40 percent were young-of-year. In any given year, less than 5 percent of all brook trout collected are greater than 178 mm or seven inches.

Rainbow trout typically live three to four years in the park, while an occasional five-year-old fish is collected. Historic data indicates annual mortality rates for rainbow trout in the park ranges from 60 to 70 percent from ages one to four. Most rainbow trout average four to 10 inches with an occasional fish reaching 14 inches. Brown trout typically live five to eight years with an occasional fish living to 12 years of age. Most brown trout average six to 14 inches with an occasional fish reaching 25 to 30 inches and eight to 10 pounds.

Acid deposition has not been directly implicated in the total loss of any brook trout populations. However, water-quality monitoring data indicate that stream pH during major storm events drops to levels known to stress brook trout.

A spatial modeling study to project southern Appalachian trout distribution in a warmer climate caused by global warming revealed a 53 percent to 97 percent loss of trout habitat by the year 2100. With increasing temperature, fragmentation would increase, leaving populations in small, isolated patches vulnerable to extirpation because of the decreased likelihood of recolonization.

In April 2006 a 30-year prohibition on catching and keeping brook trout was lifted on an experimental basis after studies concluded that fishing had little impact on the population. An environmental assessment was released on August 7, 2006, that offered two options: remove the fishing ban generally, except in newly restocked streams, or reinstate the prohibition. Park fisheries biologists consider open fishing the environmentally preferred alternative as long as a catch-and-release program can be used if the brook trout population starts to decline. The ban was lifted in 2007.

Smallmouth and Rock Bass

Smallmouth bass and rock bass inhabit the lower-most sections of park streams where the water is warmer. The smallmouth is found in numerous streams, whereas rock bass are found in only a few. During the early 1990s, these species were significantly affected by sediment and flood impacts in Abrams Creek. Studies indicated their recovery by 2000. Reduction in sediment inputs appears to be related to streambank restoration and the elimination of cattle in Cades Cove.

More in-depth information about specific species of fish found in the park can be found at http://www.discoverlifeinamerica.org/atbi/species/animals/vertebrates/fish.

Amphibians

The Smokies has the distinction of having the most diverse salamander population anywhere in the world. There are 31 species of salamanders and 12 species of frogs and toads known historically from the park. One species, the green salamander (*Aneides aeneus*), is known from only a single individual taken from beneath a log near Cherokee Orchard in 1929. Extensive searching during the past 79 years has failed to yield any other individuals. In other areas of its range, the green salamander is found in rock crevices in damp and shaded cliffs and rock outcrops as well as beneath the bark of trees. The northern cricket frog (*Acris crepitans*) is known from only four specimens taken in 1940 near the town of Chilhowee, which is now submerged beneath Lake Chilhowee just outside the western boundary of the park. Whether this species ever occurred within the park is questionable.

Erin Hyde and Ted Simons showed that salamander populations are more abundant and salamander communities are more diverse on undisturbed sites compared to mature second-growth sites in the park. They also found that several salamander species (*Desmognathus imitator, D. ocoee, D. wrighti,* and *Plethodon jordani*) showed strong positive associations (were significantly more abundant) with undisturbed sites, while members of the *Plethodon glutinosus* complex (*P. glutinosus* and *P. oconaluftee*) showed a significant association with disturbed sites.

In addition, Larissa Bailey, Ted Simons, and Kenneth Pollock have devised sampling methods to estimate site occupancy, species detection probability parameters, and temporary emigration for plethodontid salamanders. The goal is to estimate the proportion of sites that are occupied, knowing the species is not always detected, even when present. Such estimates would allow researchers to establish reliable baseline data for multiple species, compare species-specific site occupancy over time or among different studies from various regions, and identify habitat characteristics important for species presence (occurrence). One of their studies was the first to formally estimate temporary emigration (movement of an individual down into the soil) in terrestrial salamander populations, and their results verified that significant proportions of terrestrial salamander populations are subterranean. Temporary emigration was higher on low-elevation/disturbed/deciduous sites than high-elevation/undisturbed/deciduous sites since older, more mature forests have less daily and seasonal microhabitat variability than younger forests. Salamanders in moist, stable habitats would be expected to emigrate below the surface less often than salamanders found in areas with constantly changing microhabitat conditions.

Checklist of Park Salamanders:
Family Ambystomatidae—Mole salamanders (3)
    Marbled salamander (*Ambystoma opacum*)
    Mole salamander (*Ambystoma talpoideum*)
    Spotted salamander (*Ambystoma maculatum*)
Family Cryptobranchidae—Hellbender (1)
    Eastern hellbender (*Cryptobranchus alleganiensis*)
Family Necturidae—Mudpuppy (1)
    Common mudpuppy (*Necturus maculosis*)

The rather chunky marbled salamander has white or silvery crossbars on its dorsal surface. The crossbands are variable and may not always be complete.

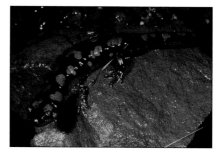

The spotted salamander can be recognized by an irregular row of round dorsolateral yellow or orange spots from the eye to the tip of the tail.

A thoroughly aquatic salamander, the hellbender has a flattened head, and each side of its body consists of a wrinkled, fleshy fold of skin. Hellbenders forage on the bottom of the water at night and feed on crayfish, worms, and aquatic insects.

The long-tailed salamander is the only yellowish salamander with vertical black markings on the tail. It is found in and under rotting logs, beneath rocks, and often in caves.

The dark herringbone pattern down the center of the back is distinctive of the pigmy salamander, the smallest salamander in the park.

Family Plethodontidae—Lungless salamanders (25)
    Black-bellied salamander (*Desmognathus quadramaculatus*)
    Black-chinned red salamander (*Pseudotriton ruber*)
    Blue Ridge spring salamander (*Gyrinophilus porphyriticus*)
    Blue Ridge two-lined salamander (*Eurycea wilderae*)
    Cave salamander (*Eurycea lucifuga*)
    Four-toed salamander (*Hemidactylium scutatum*)
    Green salamander (*Aneides aeneus*)
    Imitator salamander (*Desmognathus imitator*)
    Jordan's salamander (*Plethodon jordani*)
    Junaluska salamander (*Eurycea junaluska*)
    Longtailed salamander (*Eurycea longicauda*)
    Midland mud salamander (*Pseudotriton montanus*)
    Northern slimy salamander (*Plethodon glutinosus*)
    Ocoee salamander (*Desmognathus ocoee*)
    Pigmy salamander (*Desmognathus wrighti*)
    Santeetlah dusky salamander (*Desmognathus santeetlah*)
    Seal salamander (*Desmognathus monticola*)
    Seepage salamander (*Desmognathus aenus*)
    Shovel-nosed salamander (*Desmognathus marmoratus*)
    Southern Appalachian salamander (*Plethodon oconaluftee*)
    Southern gray-cheeked salamander (*Plethodon metcalfi*)
    Southern red-backed salamander (*Plethodon serratus*)
    Southern zigzag salamander (*Plethodon ventralis*)
    Spotted dusky salamander (*Desmognathus conanti*)
    Three-lined salamander (*Eurycea guttolineata*)
Family Salamandridae—Newts (1)
    Eastern red-spotted newt (*Notophthalmus viridescens*)

Dusky salamander climbing the face of Hen Wallow Falls. Photo by Steve Bohleber.

Hen Wallow Falls near Cosby, the destination for many of my naturalist-led hikes. Photo by Steve Bohleber.

Spotted, Marbled, and Mole Salamanders

Through most of the year it is almost impossible to find a spotted salamander in the park. This species, along with mole and marbled salamanders, belong to a family (Ambystomatidae) known as mole salamanders since they spend most of their life underground. With the first warm, springlike rains (late January to early February), spotted salamanders congregate in large numbers in a few pools and ponds. It is an unforgettable sight to see so many of these beautiful animals in one place, especially on a rainy night when illuminated by the beams of a headlamp or flashlight. There they lay their tapioca-like masses of eggs. Within a day or two the adults are gone from the pond. By mid-March the eggs have hatched into larvae, and by June the larvae have become adults. Then they leave the breeding pools and "disappear" until the following spring.

Unlike most amphibians that breed in the spring, female marbled salamanders breed and deposit their eggs in a depression under a rock or log in the fall of the year. The female guards her eggs through the winter. When late winter and early spring rains flood the breeding site, the eggs hatch into larvae and the adults leave. By the time that other amphibians are just laying their eggs, the marbled salamander larvae are well established and often feed on the eggs and young larvae of the other amphibians.

The mole salamander was not discovered in the park until 1998 and is known only from Gum Swamp in Cades Cove. Breeding probably occurs in early spring, although specific data from the park is lacking.

The skin of all members of the genus *Ambystoma* contains noxious secretions that help deter attacks from birds and from small mammals such as shrews that may share burrows with these salamanders.

Hellbender

The hellbender is the largest salamander in North America with adults reaching a length of 24 to 30 inches. Its nearest relative is the Giant Salamander of Japan that may reach a length of five feet. Hellbenders live beneath submerged rocks and boulders in the larger waterways in the park. The largest specimens recorded from the park have been a female measuring 29.125 inches in total length and a male measuring 26 inches. Both were taken in the Little Pigeon River.

In 2000 Max Nickerson of the University of Florida began conducting research on this species in the Little River. Individuals were marked with Passive Integrated Transponder (PIT) tags so that their population density and structure could be studied. Michael Freake of Lee University in Cleveland, Tennessee, assumed leadership in 2004 and is currently heading up the project. Although he has found hellbenders in Abrams Creek, at Deep Creek, and in the Oconaluftee River, the population in the Little River appears to be the largest and most successfully reproducing population. Animals of all size ranges from larvae to adults have been marked, and many have been recaptured.

Lungless Salamanders

The lungless salamanders (Plethodontidae), which constitute about 70 percent of existing salamander species, breathe through their moist, well-vascularized skin (cutaneous gas exchange) and through the lining of their mouth and pharynx (buccopharyngeal gas exchange). Lack of lungs is thought to be an adaptation to life in and around mountain streams where buoyant lungs would be a disadvantage in fast-moving, well-oxygenated waters. The terrestrial species of lungless salamanders generally restrict their above-ground foraging activity to moist or rainy conditions at night, which limits the possibility of desiccation and loss of cutaneous gas exchange.

One of the lungless salamanders is the Jordan's, or red-cheeked, salamander whose entire range is restricted to the Great Smoky Mountains National Park. Except for the red cheeks, adults are blackish and average about four inches in length. It is a terrestrial species, living beneath logs and rocks above 2,400 feet elevation. I have found these salamanders at many sites from Chimneys picnic area to the summit of Clingmans Dome. In June 1982 an estimated 1,000 to 2,000 red-cheeked salamanders were observed by two park rangers along Clingmans Dome Road. Although mass salamander movements are not unusual in some species, it had never before been reported for red-cheeked salamanders. Such movements occur as a result of summer storm activity and air temperature. Another lungless salamander, the imitator salamander, is confined to the Smokies and has many individuals with red to yellowish cheeks. This has been shown to be a classic example of Batesian mimicry

The red-cheeked salamander is unmistakable with its bright red cheek patches. Eggs are laid in small clusters inside or under damp logs or in moss. Complete development takes place within the egg; there is no aquatic larval stage as is typical with most other salamanders.

in which a palatable species (imitator) mimics a highly distasteful species (Jordan's) that is avoided by most predators. Predators learn to avoid the distasteful species as well as the similarly colored palatable species.

The smallest salamander in the park is the pigmy salamander. Adults are only between 1.5 and 2 inches in total length and have a pattern of dark chevronlike markings on their reddish-bronze backs. They are found in wet moss, under rocks and logs, and in decaying wood at moderate to high elevations (above 2,400 ft.), especially in spruce-fir forests. Females brood their tiny eggs in seepage areas; there is no aquatic larval stage. The fate of the pigmy salamander is closely tied to the spruce-fir forest. During the cooler and more moist Ice Age, the spruce-fir forest was probably continuous over the higher mountains. With the advent of warmer and drier times, the spruce and fir have been replaced by hardwood trees in the lower elevations, isolating the spruce-fir and pigmy salamanders in "islands" on and around the highest peaks. In the future, pigmy salamanders in the park may be adversely affected by the balsam woolly adelgid and by changing climatic conditions caused by global warming (see chapters 11 and 12).

Eastern Red-spotted Newt

The eastern red-spotted newt has the most complex life cycle of any amphibian in the Smokies. This species may metamorphose into an adult by passing through two "larval" stages. Eggs hatch into aquatic larvae. After several months, the aquatic larva metamorphoses into a terrestrial "larval," or juvenile, stage known as a red eft. Red efts have a rough, dry, granular skin that contains toxic chemicals which prevent predation by most vertebrates. They are bright reddish orange (a warning coloration) with two rows of black-rimmed red spots on their back. This carnivorous stage develops lungs, leaves the pond, and lives a terrestrial existence for several years. I usually find several each summer under pieces of bark, beneath leaves, or just walking on the surface of the ground. After several years, the eft returns to water, develops fins on the dorsal and ventral surfaces of its tail, and lives the remainder of its adult life as a lung-breathing, carnivorous, aquatic salamander. In areas where ponds dry up during the summer, as they often do in Cades Cove, the dull drab-green adults develop a dry, almost leathery skin and move to terrestrial refugia near the ponds. They reenter the water soon after heavy rains.

Checklist of Park Frogs and Toads:
    Family Bufonidae—True toads (2)
        Eastern American toad (*Bufo americanus*)
        Fowler's toad (*Bufo fowleri*)
    Family Hylidae—Tree Frogs (3)
        Cope's gray treefrog (*Hyla chrysoscelis*)
        Spring peeper (*Pseudacris crucifer*)
        Upland chorus frog (*Pseudacris feriarum*)
    Family Microhylidae—Narrow-mouthed toads (1)
        Eastern narrow-mouthed toad (*Gastrophryne carolinensis*)
    Family Pelobatidae—Spadefoot toads (1)
        Eastern spadefoot (*Scaphiopus holbrooki*)

One of only two true toads in the park, the American toad is covered with warts that may vary from yellowish to dark brown. It is differentiated from Fowler's toad by having only one or two large warts in each of the largest dark spots and by having the chest and forward part of the abdomen usually spotted with dark pigment.

A highly variable species, the green frog may be more brownish than green. A pair of dorsolateral ridges (folds) extend most of the length of the body.

Family Ranidae—True frogs (5)
American bullfrog (*Rana catesbeiana*)
Northern green frog (*Rana clamitans*)
Leopard frog [Experts unclear about which species inhabits the park. Either Northern Leopard Frog (*Rana pipiens*) or Southern Leopard Frog [*Rana sphenocephala*)]
Pickerel frog (*Rana palustris*)
Wood frog (*Rana sylvatica*)

True Toads (Family Bufonidae)

Toads in this family have a dry, warty skin, and they hop, as opposed to most frogs, which have a moist, relatively smooth skin, and leap. Although the two members of this family in the park superficially resemble each other, they are rather easy to differentiate. The American toad possesses just one or two warts in each dark spot on its dorsal surface, and it has a heavily pigmented chest. Fowler's toad usually has three or more warts per dark spot and usually has a virtually unspotted chest. Within the park, the American toad is much more abundant and widespread than Fowler's toad and may occur almost to the highest elevations. Fowler's toad occurs primarily in the lowlands and is rarely found above 3,000 feet.

A person does not get warts from touching a toad, but their skin-gland secretions are irritating to the eyes and mucous membranes in the mouth. Secretions by toads in this family may contain adrenaline, noradrenaline, and steroids such as bufogenine or bufotoxin. The pharmacological effect of bufotoxin resembles the effect produced by products containing digitalis and is used in treating human heart ailments.

Treefrogs (Family Hylidae)
These are small frogs with big voices. Although all members of this family possess adhesive toe discs, they are best developed in the gray treefrog. It is the most arboreal of the park's frogs. Except during the breeding season, it spends most of its time in trees and shrubs. The spring peeper and the upland chorus frog may become active as early as late February at low elevations. The high-pitched "peep" of the spring peeper is one of the earliest signs of spring. Although many people have heard these frogs, very few have ever seen them. These species breed in small woodland pools or wetlands at low elevations in Cades Cove and along the periphery on the Tennessee side of the park. After the breeding season, individuals may disperse and move to elevations as high as 4,500 to 5,000 feet.

Narrow-mouthed Toad (Family Microhylidae)
These amphibians are not true toads but are members of an entirely different family of frogs. They are small and plump with short limbs and pointed heads. A fold of skin runs across the head behind the eyes. It can be moved forward to clear debris from the eyes. These frogs are very secretive by day, hiding beneath logs and rocks. They are primarily active on damp, warm nights. The high-pitched distinctive call sounds like the bleating ("baaaa") of a lamb. When in the midst of a chorus of these frogs, one has the feeling that they are in the midst of a herd of sheep. Another unique characteristic of this frog is its method of escape. Most frogs escape by leaping. The narrow-mouthed toad escapes by running interspersed with short hops of an inch or two. This species is known from only a few localities in the western portion of the park.

True Frogs (Family Ranidae)
All of the true frogs in the park (bullfrog, green, leopard, pickerel, and wood) generally occur at elevations below 2,000 feet. The bullfrog, a resident of permanent bodies of still water, is the largest frog in the park with adults measuring up to eight inches in body length. Besides its size, this species can be distinguished from the similar-appearing green frog by the lack of skin folds (dorsolateral folds) along each side of the back.

The greenish to brown leopard frog and the light brown pickerel frog are both spotted. Leopard frogs have randomly arranged rounded dark spots on their upper surfaces, whereas the spots are squarish and are arranged in two rows in the pickerel frog. In addition, the concealed surfaces of the hind legs of the pickerel frog are bright orange or yellow. The pickerel frog is known from numerous low-elevation localities within the park, while the leopard frog is extremely rare.

The attractive wood frog has an overall body color varying from a light pinkish-tan to a very dark, almost blackish-brown. The dark eye mask is a key identifying character. While most frogs live in and around water, the wood frog spends most of its time on the dry forest floor. During the summer, it may be found in moist woodlands far from water. Unlike most other frogs, it hibernates on land

The wood frog has a call that sounds somewhat like the quack of a duck. There are a pair of lateral vocal sacs that expand when the frog is calling.

under leaves, logs, and other litter. It is one of several species that are freeze-tolerant and can survive temperatures below 0°C. These species have evolved a tolerance to slow freezing by generating increased blood glucose levels as a cryoprotectant. In addition, striated muscle function remains intact at below-freezing temperatures, its cardiac function remains nearly unchanged, and its organs undergo dehydration, presumably to prevent mechanical injury during freezing. Only for a brief period in late winter (January–February) does the wood frog return to a small pond or pool to mate and lay eggs. The breeding frenzy is something to behold. I have seen as many as eight to 10 males clustered together, all attempting to copulate with a single female.

Eastern Spadefoot Toad (Family Pelobatidae)
This species is known only from Gum Swamp in Cades Cove. It was first discovered in the park in 1999.

Mortality
Significant amphibian die-offs occurred during the spring of 1999, 2000, and 2001 in Gourley Pond in Cades Cove. Sick and recently dead wood frog tadpoles, eastern red-spotted newts, and spotted salamanders were submitted to the National Wildlife Health Center in Madison, Wisconsin, for pathological examination. The report sent to the park stated that a group of viruses known as iridoviruses were believed responsible for the die-offs. Iridovirus was isolated from the newts and spotted salamanders, and although not isolated from the wood frog tadpoles, was consistent with histological abnormalities seen in the tadpoles. This was the first time in over 25 years that iridovirus had been isolated from newts. Iridoviruses are not known to infect humans or other homiothermic (warm-blooded) species. The Health Center specifically checked for evidence of the chytrid fungus that has been

implicated in the massive, worldwide amphibian die-offs and species extinctions in Central America, but no evidence was found. Even though the pond dries up in the summers, it is thought that one or more amphibians might be carriers of this virus and that monitoring should continue. A more detailed description of these events can be found in Dodd's *The Amphibians of Great Smoky Mountains National Park* (University of Tennessee Press, 2004).

More in-depth information about specific species of amphibians can be found at http://www.discoverlifeinamerica.org/atbi/species/animals/vertebrates/amphibians.

Reptiles
There are 40 species of reptiles (turtles, lizards, and snakes) in the park.

Checklist of Park Turtles
Family Chelydridae—Snapping turtles (1)
    Common snapping turtle (*Chelydra serpentina*)
Family Emydidae—Painted, map, box, and slider turtles (4)
    Common map turtle (*Graptemys geographica*)
    Cumberland slider turtle (*Trachemys scripta*)
    Eastern box turtle (*Terrapene carolina*)
    Eastern painted turtle (*Chrysemys picta*)
Family Kinosternidae—Mud and musk turtles (2)
    Common musk turtle (*Sternotherus odoratus*)
    Stripeneck musk turtle (*Sternotherus minor*)
Family Trionychidae—Soft-shelled turtles (1)
    Eastern spiny soft-shelled turtle (*Apalone spinifera*)

Male eastern box turtles have bright red eyes; females have yellow-brown eyes. These turtles are especially active in the morning or after a rain. Young are mostly carnivorous (earthworms, snails, insect larvae); adults are mostly herbivorous (berries, fungi, fruit). Photo by Steve Bohleber.

Seven of the eight species of turtles found in the park are aquatic or semi-aquatic. The largest of these is the common snapping turtle, which is usually found in ponds or slow-moving streams below 2,500 feet. The eastern box turtle is a mostly terrestrial species with a high-domed carapace (dorsal shell) and a hinged plastron (ventral shell). It is able to withdraw its body completely inside its shell for protection. The range of the box turtle was always thought to be rare above 4,000 feet. However, studies by Dr. Ben Cash of Maryville College indicate that this species is readily found as high as 5,500 feet and is distributed throughout the park.

Checklist of Park Lizards:
    Family Anguidae—Glass lizards (1)
        Eastern slender glass lizard (*Ophisaurus attenuatus*)
    Family Phrynosomatidae—Horned lizards (1)
        Northern fence lizard (*Sceloporus undulatus*)
    Family Polychrotidae—Anoles (1)
        Northern green anole (*Anolis carolinensis*)
    Family Scincidae—Skinks (5)
        Broadheaded skink (*Eumeces* [*Plestiodon*] *laticeps*)
        Coal skink (*Eumeces* [*Plestiodon*] *anthracinus*)
        Five-lined skink (*Eumeces* [*Plestiodon*] *fasciatus*)
        Southeastern five-lined skink (*Eumeces* [*Plestiodon*] *inexpectatus*)
        Ground skink (*Scincella lateralis*)
    Family Teiidae—Whiptail lizards (1)
        Six-lined racerunner (*Cnemidophorus* [*Aspidocelis*] *sexlineatus*)

Five-lined skinks are smooth, shiny, and extremely quick. Like many species of lizards, they have the ability to sever their tail at its base (autotomy) in order to escape a predator. Over time, they will regenerate a new tail.

Eastern Slender Glass Lizard

The rarest lizard in the park is the eastern slender glass lizard. This legless lizard, which is also known as a "glass snake" or "joint snake," is found only in the extreme western end of the park. Like all lizards, it possesses movable eyelids and external ear openings, characteristics that snakes do not possess. As is the case in many lizards, especially skinks, this species has the ability to detach its tail as a defensive behavior. The tail may even break into several pieces as the glass lizard seeks protective cover. There is no truth to the mistaken belief that the broken pieces may become rejoined or may grow into entirely new individuals. A lizard that loses its tail will gradually grow a new tail.

Northern Fence Lizard

The northern fence lizard, or "swift" as it is often aptly called, is a dweller of the drier oak and pine woodlands. It prefers open, sunny areas and may often be seen perched on fences, logs, rocks, or stumps. When startled, it often dashes for the nearest tree, climbs a short distance and remains motionless on the side of the tree opposite the danger. This is a spiny lizard with sharply keeled and spiny scales. Adult males have bright bluish throats and sides which play important roles in courtship behavior.

Green Anole

The dry, piney woods are the northernmost outpost of the green anole in the mountains. It is found in areas below 1,600 feet in the western end of the park. It can often be seen along the trail to Abrams Falls. Anoles are often mistakenly called "chameleons," a name that belongs to a very different group of Old World lizards. Anoles

Anoles have pads on their toes that aid in climbing. They feed primarily on insects and spiders.

change color in response to changes in temperature, light level, and their emotional state. They are normally green to blend into their green, leafy surroundings. They change to brown in response to cooler temperatures, diminished activity, and interactions with other anoles. The change usually takes 20 minutes or more to complete. Males possess a unique pink throat fan known as a "dewlap" that can be extended during courtship displays and as a part of territorial behavior. An anole's toes are expanded into adhesive pads that aid in climbing among branches and leaves.

Skinks

Skinks are smooth, shiny, very active lizards. Some, such as the coal, five-lined, and southeastern five-lined skinks, have longitudinal stripes whose position and width are important in accurately identifying the species. Some change in coloration and pattern as they grow older. For example, young five-lined skinks, southeastern five-lined skinks, and broad-headed skinks have a bright blue tail that turns grayish as they mature. The tails of all skinks break off very easily and serve to deter potential predators, especially birds. While the predator is distracted by the wriggling tail, the skink is usually able to seek protective cover. Skinks are mainly terrestrial but may occasionally climb into trees. Broad-headed skinks are found in trees more often than other species in the park. Skinks feed primarily on insects and other arthropods.

Checklist of Park Snakes:
    Family Colubridae—Nonvipers (21)
        Black rat snake (*Elaphe* [*Pantherophis*] *obsoleta*)
        Corn snake (*Elaphe* [*Pantherophis*] *guttata*)
        Eastern garter snake (*Thamnophis sirtalis*)
        Eastern hognose snake (*Heterodon platirhinos*)
        Eastern kingsnake (*Lampropeltis getula getula*)
        Eastern black kingsnake (*Lampropeltis getula nigra*)
        Eastern milk snake (*Lampropeltis triangulum*)
        Eastern smooth earth snake (*Virginia valeriae*)
        Eastern worm snake (*Carphophis amoenus*)
        Mole kingsnake (*Lampropeltis calligaster*)
        Northern black racer (*Coluber constrictor*)
        Northern brown snake (*Storeria dekayi*)
        Northern pine snake (*Pituophis melanoleucus*)
        Northern redbelly snake (*Storeria occipitomaculata*)
        Northern ringneck snake (*Diadophis punctatus*)
        Northern scarlet snake (*Cemophora coccinea*)
        Northern water snake (*Nerodia sipedon*)
        Queen snake (*Regina septemvittata*)
        Rough green snake (*Opheodrys aestivus*)
        Scarlet kingsnake (*Lampropeltis triangulum elapsoides*)

Eastern garter snakes are extremely variable in their coloration. Unlike most other snakes, they do not lay eggs but give birth to living young.

    Southeastern crowned snake (*Tantilla coronata*)
Family Viperidae—Pit vipers (2)
    Northern copperhead (*Agkistrodon contortrix*)
    Timber rattlesnake (*Crotalus horridus*)

Two families of snakes occur in the park. The family Colubridae is the largest family of snakes and contains over 70 percent of the world's species of snakes. All species within this family in the park are nonvenomous. The family Viperidae contains the venomous pit-vipers, each of which possess a pair of heat-sensing pits posterior to their nasal openings.

The teeth of most snakes are recurved and point backwards toward the rear of the mouth, enabling the snake to better grasp and hold their prey while swallowing. The right and left sides of many snakes' mandibles (lower jaws) (as well as those of some lizards) are joined only by an elastic ligament so that the symphysis (joint) can stretch to accommodate large prey. Further expansion is possible because both halves of the lower jaw are suspended from the skull by a long, folding strut which is divided into two hinged sections similar to a folding carpenter's rule. As large prey is engulfed, these joints swing outward on their flexible struts, greatly increasing the diameter of the throat region. By alternately engaging and disengaging the recurved teeth on each side of the mouth, prey is gradually drawn into the oral cavity and passed to the pharynx and esophagus.

Nonvenomous Snakes

Several of the park's snakes, including the black rat snake, corn snake, and eastern kingsnake, are constrictors. They stun their prey with a lightninglike strike to the head and then immediately enfold the victim within the tight coils of their body and squeeze the prey until it suffocates. Although all three species feed on mice and other small mammals, the eastern kingsnake often feeds on other snakes

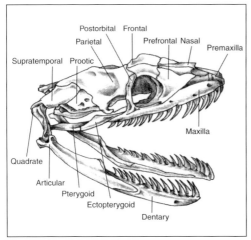

Lateral view of python skull. The right and left sides of the mandible (lower jaw) are joined only by an elastic ligament so that the symphysis (joint) can stretch. In addition, both halves of the lower jaw are suspended from the skull by a long, folding strut, which is divided into two hinged sections. As large prey is engulfed, these joints swing outward on their flexible struts, thus increasing the diameter of the throat region. Recurved teeth securely hold the prey as upper and lower bones on one side of the jaw slide forward and backward alternately with bones of the other side. From *Vertebrate Biology* by D. W. Linzey (New York: McGraw-Hill, 2001).

including rattlesnakes and copperheads. The northern watersnake is a favorite victim and may account for the kingsnake's fondness for stream margins. Both the black rat snake and corn snake are adept tree climbers, which allows them to also feed on birds and their eggs. The black rat snake may exceed six feet in length, making it the largest snake in the park.

Rat snakes have a tendency to vibrate their tails rapidly when alarmed. I will never forget the morning in 1964 when I was checking my golden mouse live traps and came upon one that was closed. Thinking I had captured one of my resident animals, I picked up the trap so that I could remove the animal and identify it. Much to my surprise, as I picked the trap up, whatever was inside suddenly began

Often called the "red rat snake," the corn snake is a beautiful red or orange snake with a checkered belly. A stripe with dark borders extends backward from the eye onto the neck.

The northern water snake is harmless and semiaquatic. It is the only large water snake in the park and feeds on frogs, salamanders, fish, and crayfish.

rattling. I immediately set the unopened trap down, thinking that it contained a small timber rattlesnake. After contemplating the situation for several minutes, I devised a method of opening the trap without exposing my fingers. To my surprise, a young black rat snake emerged and crawled away. It must have been as happy to be released as I was to not have to deal with a rattlesnake!

The nonvenomous northern water snake, which is common in the rivers and streams throughout the park, is often mistaken for a copperhead or rattlesnake. It is also often mistakenly called a water moccasin, a venomous species that does not occur in eastern Tennessee or western North Carolina. The northern water snake, which is one of the most abundant snakes in the park, is usually aggressive when disturbed.

Venomous Snakes

Only two poisonous (venomous) snakes occur in the park—the northern copperhead and the timber rattlesnake. Copperheads have a coppery-colored head and a brownish or brownish-red hourglass pattern (often broken or discontinuous) on the dorsum. Young copperheads have a bright yellow tail tip. Copperheads prefer dry, rocky woodlands on south-facing slopes and ridge crests, although I have almost stepped on one in Cades Cove near the edge of a temporary pond and have had my hand come too close for comfort to two others while checking small mammal live traps at Cosby and at Tremont. The ground color of timber rattlesnakes may range from yellow to brown or even black. The presence of a rattle is an obvious indicator, but individuals can lose their rattle. Rattlesnakes occur throughout the park but are found most often on steep rocky slopes and in the dense grass of the balds. No fatalities from snakebites have ever been recorded in the park.

Copperheads are quiet, well-camouflaged snakes. They are gregarious, especially in autumn, when they gather at denning or hibernating sites, often with other species of snakes. They feed primarily on mice but also take small birds, frogs, and insects.

Often called the "banded" or "velvet-tail rattler," the timber rattlesnake occurs in two major color patterns—a yellow phase and a black phase. They may congregate in dens to hibernate along with copperheads and other snakes.

One of the walks that I led from the Cosby Campground was to Sutton Ridge, an overlook along the Lower Cammerer Trail. On the side trail going up the ridge, I could almost always count on seeing one or more copperheads sunning themselves on the dry hillside. I would alert members of my party ahead of time that they might have an opportunity to get a photograph but to be careful. When we reached the side trail, I would tell everyone to remain on the main trail until I had an opportunity to see if a copperhead was present. If one was present, I would bring the hikers up the trail single-file and point out the snake's location so that it could be photographed. We never experienced any problem, and many of the hikers thanked me for giving them an opportunity to see something that they had never seen before.

Both the copperhead and the timber rattlesnake are pit-vipers, meaning that they have a pair of deep facial heat-sensing pits on each side of the head between the eye and nostril. These pits are directed forward and are covered by a thin trans-

The heat-sensing facial pits on either side of the head are sensory organs that help the copperhead accurately strike its prey.

Table 6.1. Diet of rattlesnakes and copperheads in Great Smoky Mountains National Park.

	Rattlesnake	Copperhead
Cicada	0	29
Unidentified insects	0	2
Salamanders	0	1
Snakes	0	2
Lizards	0	2
Birds	3	1
Shrews and moles	6	7
Weasels	1	0
Chipmunks	4	0
Squirrels (*Sciurus* sp.)	5	0
Southern flying squirrels	1	0
Peromyscus sp.	21	0
Other mice (*Clethrionomys, Microtus chrotorrhinus, Microtus pinetorum, Napaeozapus*)	18	5
Rabbit	3	0
Unidentified small mammals	6	1
Total	63	50

parent membrane. These specialized pits house infrared receptors that respond to radiant heat. Thus, they can detect the presence of objects, including homiothermic animals on which these snakes prey, even if the object is only slightly warmer than the environment. Because of this extreme sensitivity, these snakes, which are at least partly nocturnal, can locate and strike accurately at objects in the dark. They prey primarily on small mammals that are mainly nocturnal and whose temperatures usually differ from those of their surroundings.

Tom Savage, a seasonal naturalist, carried out a study of the food habits of copperheads and rattlesnakes in the park in 1963–64 (Table 6.1). Copperheads preyed on both homiothermic (warm-blooded) and ectothermic (cold-blooded) animals; the rattlesnakes preyed only on homiothermic animals. Of the total small mammals (59), rattlesnakes took 78 percent and copperheads, 22 percent; of the birds (4), rattlesnakes took 75 percent and copperheads, 25 percent. Small mammals appeared as an important part of the diet of the rattlesnake, much less so of the copperhead.

Savage noted that since most of the snakes were taken during July and August, when cicadas are most common, the high number of cicadas should be considered a seasonal effect. When cicadas are not available, it is expected that small mammals make up a larger percentage of the copperhead diet.

For in-depth discussions of the natural history and biology of each species, go to discoverlifeinamerica.org/atbi/species/animals/vertebrates/reptiles.

Birds

A total of 243 species of birds have been recorded in the park. Sixty-one species are considered permanent residents, and at least 110 are known to breed in the park. The following birds are considered Permanent/Breeder or Regular Migrant.

Checklist of Park Birds:
- Family Accipitridae—Hawks and allies (8)
    - Bald eagle (*Haliaeetus leucocephalus*)—breeds just outside park
    - Broad-winged hawk (*Buteo platypterus*)
    - Cooper's hawk (*Accipiter cooperii*)
    - Northern harrier (*Circus cyaneus*)
    - Osprey (*Pandion haliaetus*)
    - Red-tailed hawk (*Buteo jamaicensis*)
    - Red-shouldered hawk (*Buteo lineatus*)
    - Sharp-shinned hawk (*Accipiter striatus*)
- Family Alcedinidae—Kingfishers and allies (1)
    - Belted kingfisher (*Ceryle alcyon*)
- Family Anatidae—Ducks, geese, and swans (3)
    - Canada goose (*Branta canadensis*)
    - Mallard (*Anas platyrhynchos*)
    - Wood duck (*Aix sponsa*)
- Family Apodidae—Swifts (1)
    - Chimney swift (*Chaetura pelagica*)
- Family Ardeidae—Herons, bitterns, and egrets (3)
    - American bittern (*Botaurus lentiginosus*)
    - Great blue heron (*Ardea herodias*)
    - Green heron (*Butorides virescens*)
- Family Bombycillidae—Waxwings (1)
    - Cedar waxwing (*Bombycilla cedrorum*)
- Family Caprimulgidae—Nighthawks and nightjars (2)
    - Chuck-will's-widow (*Caprimulgus carolinensis*)
    - Whip-poor-will (*Caprimulgus vociferous*)
- Family Cardinalidae—Cardinals and allies (4)
    - Blue grosbeak (*Passerina caerulea*)
    - Indigo bunting (*Passerina cyanea*)
    - Northern cardinal (*Cardinalis cardinalis*)
    - Rose-breasted grosbeak (*Pheucticus ludovicianus*)
- Family Cathartidae—New World vultures (2)
    - Black vulture (*Coragyps atratus*)
    - Turkey vulture (*Cathartes aura*)
- Family Certhiidae—Creepers (1)
    - Brown creeper (*Certhia americana*)

Indigo buntings spend the winter months in Central and South America. They are secretive, usually seen along woodland edges feeding on insects. This species exhibits sexual dimorphism: the male is a vibrant blue while the female is a light-brown finch-like bird.

Family Charadriidae—Plovers and lapwings (1)
    Killdeer (*Charadrius vociferus*)
Family Colombidae—Pigeons and doves (1)
    Mourning dove (*Zenaida macroura*)
Family Corvidae—Jays, ravens, and crows (3)
    American crow (*Corvus brachyrhynchos*)
    Blue jay (*Cyanocitta cristata*)
    Common raven (*Corvus corax*)
Family Cuculidae—Cuckoos (2)
    Black-billed cuckoo (*Coccyzus erythropthalmus*)
    Yellow-billed cuckoo (*Coccyzus americanus*)
Family Emberizidae—New World sparrows (13)
    Chipping sparrow (*Spizella passerina*)
    Dark-eyed junco (*Junco hyemalis*)
    Eastern towhee (*Pipilo erythrophthalmus*)
    Field sparrow (*Spizella pusilia*)
    Fox sparrow (*Passerella iliaca*)
    Grasshopper sparrow (*Ammodramus savannarum*)
    Lincoln's sparrow (*Melospiza lincolnii*)
    Savannah sparrow (*Passerculus sandwichensis*)
    Song sparrow (*Melospiza melodia*)
    Swamp sparrow (*Melospiza georgiana*)
    Vesper sparrow (*Pooecetes gramineus*)
    White-crowned sparrow (*Zonotrichia leucophrys*)
    White-throated sparrow (*Zonotrichia albicollis*)
Family Falconidae—Falcons and caracaras (2)
    American kestrel (*Falco sparverius*)
    Peregrine falcon (*Falco peregrinus*)
Family Fringillidae—Finches and allies (6)
    American goldfinch (*Carduelis tristis*)
    Evening grosbeak (*Coccothraustes vespertinus*)

House finch (*Carpodacus mexicanus*)
Pine siskin (*Carduelis pinus*)
Purple finch (*Carpodacus purpureus*)
Red crossbill (*Loxia curvirostra*)
Family Hirundinidae—Swallows and martins (3)
    Barn swallow (*Hirundo rustica*)
    Northern rough-winged swallow (*Stelgidopteryx serripennis*)
    Tree swallow (*Tachycineta bicolor*)
Family Icteridae—Blackbirds, cowbirds, orioles (8)
    Baltimore oriole (*Icterus galbula*)
    Bobolink (*Agelaius phoeniceus*)
    Brown-headed cowbird (*Molothrus ater*)
    Common grackle (*Quiscalus quiscula*)
    Eastern meadowlark (*Sturnella magna*)
    Orchard oriole (*Icterus spurius*)
    Red-winged blackbird (*Dolichonyx oryzivorus*)
    Rusty blackbird (*Euphagus carolinus*)
Family Mimidae—Catbirds, mockingbirds, and thrashers (3)
    Brown thrasher (*Toxostoma rufum*)
    Gray catbird (*Dumetella carolinensis*)
    Northern mockingbird (*Mimus polyglottos*)
Family Motacillidae—Wagtails and pipits (1)
    American pipit (*Anthus rubescens*)
Family Odontophoridae—New World quail (1)
    Northern bobwhite (*Colinus virginianus*)
Family Paridae—Titmice and chickadees (3)
    Black-capped chickadee (*Poecile atricapillus*)
    Carolina chickadee (*Poecile carolinensis*)
    Tufted titmouse (*Baeolophus bicolor*)
Family Parulidae—Wood-warblers (30)
    American redstart (*Setophaga ruticilla*)
    Black-and-white warbler (*Mniotilta varia*)
    Blackburnian warbler (*Dendroica fusca*)
    Blackpoll warbler (*Dendroica striata*)
    Black-throated blue warbler (*Dendroica caerulescens*)
    Black-throated green warbler (*Dendroica virens*)
    Canada warbler (*Wilsonia canadensis*)
    Cape May warbler (*Dendroica tigrina*)
    Cerulean warbler (*Dendroica cerulea*)
    Chestnut-sided warbler (*Dendroica pensylvanica*)
    Common yellowthroat (*Geothlypis trichas*)
    Golden-winged warbler (*Vermivora chrysoptera*)

  Hooded warbler (*Wilsonia citrina*)
  Kentucky warbler (*Oporornis formosus*)
  Louisiana waterthrush (*Seiurus motacilla*)
  Magnolia warbler (*Dendroica magnolia*)
  Nashville warbler (*Verivora ruficapilla*)
  Northern parula (*Parula americana*)
  Northern waterthrush (*Seiurus noveboracensis*)
  Ovenbird (*Seiurus aurocapillus*)
  Palm warbler (*Dendroica palmarum*)
  Pine warbler (*Dendroica pinus*)
  Prairie warbler (*Dendroica discolor*)
  Swainson's warbler (*Limnothlypis swainsoni*)
  Tennessee warbler (*Vermivora peregrina*)
  Worm-eating warbler (*Helmitheros vermivorus*)
  Yellow-breasted chat (*Icteria virens*)
  Yellow-rumped warbler (*Dendroica coronata*)
  Yellow-throated warbler (*Dendroica dominica*)
  Yellow warbler (*Dendroica petechia*)
Family Phasianidae—Domestic fowl and game birds (2)
  Ruffed grouse (*Bonasa umbellus*)
  Wild turkey (*Meleagris gallopavo*)
Family Picidae—Woodpeckers and allies (8)
  Downy woodpecker (*Picoides pubescens*)
  Hairy woodpecker (*Picoides villosus*)
  Northern flicker (*Colaptes auratus*)
  Pileated woodpecker (*Dryocopos pileatus*)

The pileated woodpecker is a relatively shy bird that prefers large tracts of forest. When looking for insects to eat, it excavates oval holes up to several feet long in tree trunks. The favorite food is carpenter ants. Photo by Steve Bohleber.

    Red-bellied woodpecker (*Melanerpes carolinus*)
    Red-cockaded woodpecker (*Picoides borealis*)
    Red-headed woodpecker (*Melanerpes erythrocephalus*)
    Yellow-bellied sapsucker (*Sphyrapicus varius*)
Family Regulidae—Kinglets (2)
    Golden-crowned kinglet (*Regulus satrapa*)
    Ruby-crowned kinglet (*Regulus calendula*)
Family Scolopacidae—Sandpipers, phalaropes, and allies (4)
    American woodcock (*Scolopax minor*)
    Common snipe (*Gallinago gallinago*)
    Solitary sandpiper (*Tringa solitaria*)
    Spotted sandpiper (*Actitis macularius*)
Family Sittidae—Nuthatches (2)
    Red-breasted nuthatch (*Sitta canadensis*)
    White-breasted nuthatch (*Sitta carolinensis*)
Family Strigidae—Typical owls (4)
    Barred owl (*Strix varia*)
    Eastern screech owl (*Megascops asio*)
    Great horned owl (*Bubo virginianus*)
    Northern saw-whet owl (*Aegolius acadicus*)
Family Sturnidae—Starlings and mynas (1)
    European starling (*Sturnus vulgaris*)
Family Sylviidae—Gnatcatchers and verdin (1)
    Blue-gray gnatcatcher (*Polioptila caerulea*)
Family Thraupidae—Tanagers (2)
    Scarlet tanager (*Piranga olivacea*)
    Summer tanager (*Piranga rubra*)
Family Trochilidae—Hummingbirds (1)
    Ruby-throated hummingbird (*Archilochus colubris*)
Family Troglodytidae—Wrens (4)
    Carolina wren (*Thryothorus ludovicianus*)
    House wren (*Troglodytes aedon*)
    Marsh wren (*Cistothorus palustris*)
    Winter wren (*Troglodytes troglodytes*)
Family Turdidae—Thrushes (7)
    American robin (*Turdus migratorius*)
    Eastern bluebird (*Sialia sialis*)
    Gray-cheeked thrush (*Catharus minimus*)
    Hermit thrush (*Catharus guttatus*)
    Swainson's thrush (*Catharus ustulatus*)
    Veery (*Catharus fuscescens*)
    Wood thrush (*Hylocichla mustelina*)

Family Tyrannidae—Flycatchers (8)
    Acadian flycatcher (*Empidonax virescens*)
    Eastern kingbird (*Tyrannus tyrannus*)
    Eastern phoebe (*Sayornis phoebe*)
    Eastern wood-pewee (*Contopus virens*)
    Great crested flycatcher (*Myiarchus crinitus*)
    Least flycatcher (*Empidonax minimus*)
    Olive-sided flycatcher (*Contopus cooperi*)
    Willow flycatcher (*Empidonax trailii*)
Family Tytonidae—Barn owl (1)
    Barn owl (*Tyto alba*)
Family Vireonidae—Vireos (4)
    Blue-headed vireo (*Vireo solitarius*)
    Red-eyed vireo (*Vireo olivaceus*)
    White-eyed vireo (*Vireo griseus*)
    Yellow-throated vireo (*Vireo flavifrons*)

Nesting

Birds in the park nest in a variety of places. Some nest high in trees, some are hole nesters, some nest in shrubs, and some nest on the ground. Nesting materials range from twigs and branches for some of the larger nests to leaves, grasses, moss, and mud for other species. Animal hair and old snake skins are two favorites that tufted titmice use to line their nests.

American Crow

American crows will eat almost anything—plant or animal, living or dead. They frequently invade the nests of other birds, taking either eggs or young. This nest predation explains the frequent sighting of crows being attacked or harassed by other birds.

American Goldfinch

Goldfinches are often referred to as "wild canaries." They are fairly common permanent residents of the park, but are much more common below 4,500 feet elevation. They are unique because they are the latest species in the park to begin breeding, with nest-building and brood rearing occurring from late July into September. This late nesting is due to their dependence on thistle as a lining for their nests and for the thistle seeds which are fed to the nestlings.

Dark-eyed Junco

Dark-eyed juncos (locals call them "snowbirds") are the most frequently seen bird in the high country. The Snowbird Mountains, a range of mountains near the Smokies, are named for them. Juncos nest on the ground and wait until the last moment to leave their nest. On more than one occasion while hiking in the high country, I have been momentarily startled by such activity at the edge of the trail! Flashing white outer tail feathers are a good field mark. Two populations of juncos live within the Great Smokies. During the winter, they are found together in the park's lower elevations. With the coming of spring, the local population moves up into the high country, while the second population moves northward some 600 to 700 miles. Thus, the local juncos reach their "northern" breeding habitat by a vertical migration of several thousand feet, while the overwintering juncos from the north migrate back and forth well over a thousand miles.

Belted Kingfisher

These birds are most frequently encountered along lower-altitude waterways in the park where they perch on branches and use their keen eyesight to look for

Often seen perched on a branch near the water, belted kingfishers dive headfirst into the water for small fish and return to a branch to eat. They regurgitate pellets of bone after meals since they are unable to pass the bones through their digestive tract.

fishes swimming in the water below. Unlike most birds in which the males are more brightly colored, the female belted kingfisher wears the brighter color. She possesses a broad rust or chestnut band across her lower breast, a band that is absent in the male. With their bills and feet, kingfishers dig burrows in the sides of steep stream banks. Within these burrows they nest and raise their young.

Brown-headed Cowbird

If anyone tells you that they have found a brown-headed cowbird's nest, they are either mistaken or they are pulling an April Fool's trick on you. Cowbirds of the New World and cuckoos of the Old World are social parasites. They both lay their eggs in nests of other bird species, often removing one egg from the host's nest prior to laying their own. Their deception involves surveillance, stealth, surprise, and speed. For example, in less than 10 seconds, the female cuckoo alights on a nest, lays her own egg, removes one host egg, and is gone. Cowbird behavior is similar. Besides having a shorter incubation period than their host species, cowbird eggs hatch before many host eggs by disrupting incubation of the smaller eggs and, possibly, hatching in response to stimuli from host eggs. In addition, young cowbirds are usually larger than the natural young in the parasitized nest, and they either take the lion's share of the food or eject the host young from the nest. In North America, 220 species have been recorded as having been parasitized by brown-headed cowbirds, with 144 species actually rearing young cowbirds. This differ-

A member of the blackbird family, the brown-headed cowbird once followed bison to feed on insects attracted to the animals.

ence in the number of species parasitized versus those actually rearing cowbirds is due to host recognition and counter-strategies: deserting the nest, rejecting the cowbird egg, or depressing the egg into the bottom of the nest. One of the brown-headed cowbird's most frequent victims in the park is the chestnut-sided warbler.

In studies done by George Farnsworth and Ted Simons in the park from 1993 to 1997, only seven of 416 (less than 2 percent) monitored wood thrush nests were parasitized. Each nest contained one brown-headed cowbird egg. Five of these seven nests fledged a total of 14 wood thrushes and four cowbird chicks, suggesting that nest parasitism in these study areas had only a negligible effect on wood thrush nesting success.

Wild turkeys are strong fliers and can approach 60 miles per hour (97 km). They have excellent hearing, and their eyesight is said to be three times better than humans.

Louisiana Waterthrush

The Louisiana waterthrush is a small bird that spends most of its time on the ground along the margins of streams. It feeds mostly on insects and spiders, but may also take an occasional snail, crayfish, or small fish. While most small birds hop from place to place on the ground, the waterthrush walks. It "bobs" or "teeters" in a very characteristic way. This habit has given rise to the nickname "wagtail." This species is the most abundant species captured at the Tremont bird banding station (see below).

Wild Turkey

The largest bird in the park is the wild turkey. These permanent residents feed on insects, seeds, and fruit. They are strong fliers that can fly straight up, then away. Their eyesight and hearing are excellent. At night, they roost in trees. Males are known as toms, females as hens, and young as chicks or poults. Populations are affected by the occasional failure of the acorn crop and chicks, in particular, by cold and flooding.

Nuthatches

If you have ever noticed the movement of birds on the trunk of a tree, you realize that just about every species moves *upward* as they search for food. Nuthatches, however, almost always move *downward*. This unique behavior can serve as a good field characteristic for identifying the two species that are permanent residents in the park: the red-breasted nuthatch and the white-breasted nuthatch.

Ruby-throated Hummingbird

Ruby-throated hummingbirds, the only birds capable of flying backward, employ a rigid wing with little of the "wrist" and "elbow" flex of other birds. The entire wing is swiveled back and forth from the shoulder as a fixed blade. As with the rotor-blade of a helicopter, it produces the bird's hovering flight.

The white-breasted nuthatch's habit of hopping headfirst down tree trunks helps it see insects and insect eggs that birds climbing up the trunk might miss. Its incredible climbing agility comes from an extra-long hind toe claw or nail, nearly twice the size of the front toe claws.

The ruby-throated hummingbird is the smallest bird in the park. It does not sing but will chatter or buzz to communicate. It constructs a nest of plant material and spider webs, gluing pieces of lichen on the outside of the nest for camouflage.

Hummingbirds typically hover to feed more than a hundred times a day, consuming about 20 percent of their daylight hours with this expensive aerobic activity. Hover-feeding, with wingbeat frequencies up to 80–100 per second, usually lasts less than a minute. Unlike most other birds, lift during hovering is generated in both up- and down-strokes in the wingbeat cycle. The main flight muscles—the pectoralis and the supracoracoideus—are highly developed and constitute about 30 percent of the body mass. Hummingbird heart rates are about 500 beats per minute at rest and may increase to about 1,300 per minute during hovering flight. Some hummingbirds, including the ruby-throated hummingbird, fly nonstop for

Hummingbirds are able to hover by beating their wings rapidly in a figure-eight pattern—first forward and down, then backward and up. Hovering enables the birds to sip nectar from flowers that are too small and fragile on which to land. Illustration by Laurie Taylor.

approximately 700 miles across the Gulf of Mexico to Central America during their annual migration, a trip that may require about 20 hours to complete. To provide sufficient fuel for their annual migratory journey, these tiny travelers may add 50 percent to their normal weight in the days before departure.

Wood Thrush

The wood thrush, which is slightly smaller than a robin, has a rust-colored head and large spots on a snowy-white breast. These birds usually nest in moist woodlands with well-developed undergrowth, especially along streams. They transform the woodlands with a melody of great beauty and power. The clear, flutelike song consists of a series of short phrases broken by rather long pauses. Singing most often occurs at dawn and dusk but may begin half an hour before daybreak.

This Neotropical migrant winters in Mexico and Central America and breeds throughout the eastern half of the United States and southern Canada. Mated pairs build their nest of grass and weed stalks five to 12 feet up in dense shrubbery in the fork of a small tree or on a horizontal limb. The inner cup has a middle layer of mud or leaf mold and a lining of rootlets. Both parents share the duties of feeding and caring for their young.

Ted Simons and his graduate students have conducted intensive studies of wood thrushes in the park since 1993. These studies have included a comparison of breeding bird community structure and composition in undisturbed old-growth forest and mature second-growth forest, nesting success, predation, evaluating the park as a population source, and developing a GIS-based habitat model to predict wood thrush presence or absence. Among their findings were that this species nested predominantly in small (less than 10 cm dbh) eastern hemlocks (*Tsuga canadensis*) (336 out of 400 nests; 84 percent) and in rhododendron (*Rhododendron maximum*) (45 out of 400 nests; 11.3 percent). Those in hemlocks were generally surrounded by many other small hemlocks. Successful nesting was significantly correlated with nest concealment. There was a relatively high productivity of 3.31 nestlings per successful nest, but predation was found to be a potentially important constraint on nestlings in large contiguous forests such as Great Smoky Mountains National Park. While the park was found to clearly be functioning as a substantial local population source for wood thrushes across the southern Appalachians, its potential to sustain regional or continental wood thrush populations is limited.

Bird Banding

Spring migration usually reaches its peak during late April. Fall migration for most species occurs during September and October.

A bird banding station (technically known as a Monitoring Avian Productivity and Survivorship Station [MAPS]) has operated at the Great Smoky Mountains Institute at Tremont since 2000 (Table 6.2). A total of 28 species have been banded at Tremont since 2000. The Louisiana waterthrush has constituted the majority of the birds banded (224 out of 591 total captures; 37.9 percent).

At Purchase Knob, a total of 55 species have been recorded since 2002. The most common species have been the chestnut-sided warbler, dark-eyed junco, and ruby-throated hummingbird. These species have varied in abundance from 11 percent to 22 percent of the total number of birds banded from 2002 to 2007.

Data from both sites are sent to the Institute for Bird Populations in California as part of a continent-wide effort to monitor bird population trends. Fecal samples, ectoparasites, blood smears, and feather samples are collected from some birds for studies of parasites and population genetics.

For in-depth discussions of the natural history and biology of each species, go to discoverlifeinamerica.org/atbi/species/animals/vertebrates/birds.

Table 6.2. Bird-banding data from Tremont (2000–2007) and Purchase Knob (2002–2007). The number of captures includes both recaptures and unbanded birds.

Site	Year	Captures
Tremont	2000	84
	2001	86
	2002	86
	2003	56
	2004	73
	2005	61
	2006	78
	2007	67
Purchase Knob	2002	317
	2003	358
	2004	369
	2005	278
	2006	280
	2007	247

Mammals

A total of 70 species of mammals either currently inhabit the park or inhabited the park during historical times.

Checklist of Park Mammals:
    Family Bovidae—Bison (1)
        American bison (*Bison bison*) [extirpated from Park during late 1700s]
    Family Canidae—Wolves and foxes (5)
        Coyote (*Canis latrans*)
        Gray fox (*Urocyon cinereoargenteus*)
        Gray wolf (*Canis lupus*) [extirpated since about 1900]
        Red fox (*Vulpes vulpes*)
        Red wolf (*Canis rufus*) [see discussion under Reintroductions, chapter 8]
    Family Castoridae—Beaver (1)
        American beaver (*Castor canadensis*) [once widespread, but extirpated in late 1800s; recolonized park naturally in April 1966]
    Family Cervidae—Deer (2)
        Elk (Wapiti) (*Cervus elaphus*) [see discussion under Reintroductions, chapter 8]
        White-tailed deer (*Odocoileus virginianus*)

The gray fox is the only North American canid with true climbing ability. It occasionally forages in trees and frequently takes refuge in them, especially leaning or thickly branched ones.

Family Didelphidae—Opossums (1)
    Opossum (*Didelphis virginiana*)
Family Dipodidae—Jumping mice (2)
    Meadow jumping mouse (*Zapus hudsonius*)
    Woodland jumping mouse (*Napaeozapus insignis*)
Family Felidae—Cats (2)
    Bobcat (*Lynx rufus*)
    Mountain lion (*Puma concolor*) [see discussion under Endangered
        Species, chapter 7]
Family Leporidae—Rabbits and hares (2)
    Appalachian cottontail (*Sylvilagus obscurus*)
    Eastern cottontail (*Sylvilagus floridanus*)

The innermost toes of the hind foot of the opossum lack claws and are opposable. The long tail is prehensile.

As many as a dozen newborn opossums can fit in a teaspoon. The female's pouch contains 13 teats.

A spotted skunk is not easily intimidated. Its movements are slow and deliberate. If alarmed, it will often stomp its feet, raise its tail, stand on its front feet, and walk stiff-legged, all as a warning that it might spray.

Family Mephitidae—Skunks (2)
    Eastern spotted skunk (*Spilogale putorius*)
    Striped skunk (*Mephitis mephitis*)
Family Muridae—Murid rats and mice (17)
    Allegheny woodrat (*Neotoma magister*)
    Black rat (*Rattus rattus*)
    Cotton mouse (*Peromyscus gossypinus*)
    Deer mouse (*Peromyscus maniculatus*)
    Eastern harvest mouse (*Reithrodontomys humulis*)
    Golden mouse (*Ochrotomys nuttalli*)
    Hispid cotton rat (*Sigmodon hispidus*)
    House mouse (*Mus musculus*)
    Marsh rice rat (*Oryzomys palustris*)
    Meadow vole (*Microtus pennsylvanicus*)
    Muskrat (*Ondatra zibethicus*)
    Norway rat (*Rattus norvegicus*)
    Rock vole (*Microtus chrotorrhinus*)
    Southern bog lemming (*Synaptomys cooperi*)
    Southern red-backed vole (*Clethrionomys gapperi*)
    White-footed mouse (*Peromyscus leucopus*)
    Woodland vole (*Microtus pinetorum*)
Family Mustelidae—Otters, weasels, and mink (3)
    Long-tailed weasel (*Mustela frenata*)
    Mink (*Mustela vison*)
    Northern river otter (*Lontra canadensis*)

The mink is an excellent swimmer and spends much time foraging along streams. Prominent anal glands emit a powerful musky secretion when the animal is excited or disturbed.

Family Procyonidae—Raccoons (1)
    Raccoon (*Procyon lotor*)
Family Sciuridae—Woodchucks, chipmunks, and squirrels (7)
    Eastern chipmunk (*Tamias striatus*)
    Eastern gray squirrel (*Sciurus carolinensis*)
    Fox squirrel (*Sciurus niger*)
    Northern flying squirrel (*Glaucomys sabrinus*) [see discussion under Endangered Species, chapter 7]
    Red squirrel (*Tamiasciurus hudsonicus*)
    Southern flying squirrel (*Glaucomys volans*)
    Woodchuck (*Marmota monax*)
Family Soricidae—Shrews (8)
    Least shrew (*Cryptotis parva*)

Raccoons are omnivorous, feeding on a great variety of plant and animal foods. They are primarily nocturnal with excellent senses of hearing, smell, and night vision.

The chipmunk is a small, solitary, diurnal squirrel that lives primarily on the ground. It is, however, an expert tree climber. Seeds are transported in well-developed internal cheek pouches.

Locally known as the "mountain boomer," the red squirrel is easily recognized by its rufous color and small size, about half that of the gray squirrel. It is diurnal but particularly active at dawn and dusk.

    Long-tailed shrew (*Sorex dispar*)
    Masked shrew (*Sorex cinereus*)
    Pygmy shrew (*Sorex hoyi*)
    Short-tailed shrew (*Blarina brevicauda*)
    Smoky shrew (*Sorex fumeus*)
    Southeastern shrew (*Sorex longirostris*)
    Water shrew (*Sorex palustris*)
Family Suidae—Pigs (1)
    European wild hog (*Sus scrofa*)
Family Talpidae—Moles (3)
    Eastern mole (*Scalopus aquaticus*)
    Hairy-tailed mole (*Parascalops breweri*)
    Star-nosed mole (*Condylura cristata*)
Family Ursidae—Bears (1)
    Black bear (*Ursus americanus*)
Family Vespertilionidae—Bats (11)
    Big brown bat (*Eptesicus fuscus*)
    Eastern pipistrelle (*Pipistrellus subflavus*)
    Eastern red bat (*Lasiurus borealis*)
    Eastern small-footed bat (*Myotis leibii*)
    Evening bat (*Nycticeius humeralis*)
    Hoary bat (*Lasiurus cinereus*)
    Indiana bat (*Myotis sodalis*) [see discussion under Endangered Species, chapter 7]
    Little brown bat (*Myotis lucifugus*)
    Northern long-eared bat (*Myotis septentrionalis*)
    Rafinesque's big-eared bat (*Corynorhinus rafinesquii*)
    Silver-haired bat (*Lasionycteris noctivagans*)

Coyotes

Coyotes originally inhabited portions of western North America. As forests were cleared, however, their range in the United States expanded eastward. The clearing of forest land created a lot of "edge effect," where the fields go up to fence rows that increase the habitat for rodents and rabbits. Many coyotes were liberated by fox hunters in the southern states who had similar-appearing coyote pups shipped to them instead of fox pups. In addition, the U.S. Fish and Wildlife Service has documented 20 different points in the southeastern United States where coyotes were released by people who planned to run them with hounds. Three of the known releases were in Tennessee–Hardeman County prior to 1930; Hickman County

Coyotes are mainly nocturnal and active during all seasons. Their senses of sight, hearing, and smell are well developed. They may travel separately or in family units.

about 1935; and Sequatchie County, date unknown. The majority of the releases were carried out by fox hunters who were looking for something different to run with their dogs. Coyotes were first observed in the park in Cades Cove by Charles Remus on June 6, 1982, and now occur throughout the park. They prefer open woodlands, woodland borders, and brushy areas and are often seen by park visitors. Their primary foods consist of rabbits, rodents, and carrion. They feed on deer carcasses and are responsible for killing newborn fawns in Cades Cove and possibly elk calves in Cataloochee. Wherever coyotes establish themselves in large numbers, red foxes disappear, in contrast to the gray fox which coexists with coyotes.

Park Service management policy states that native animals in the parks will be perpetuated for their essential role in the natural ecosystem. In addition, the parks will strive to maintain the natural abundance, behavior, diversity, and ecological integrity of native animals in natural portions of the parks as part of the ecosystem. The policy statement sums up the status of the coyote in the Smokies by defining "native": "Native species are those that occur, *or occurred due to natural processes* on those lands designated as the park." These do not include species that have moved into those areas, directly or indirectly as the result of human activities. The coyote, therefore, under current policy guidelines is considered to be naturalized and a part of the park's fauna and enjoys the protection status afforded all native species in the park.

American Beaver

Beaver were once widespread throughout the lower elevations of the Smokies, but they were extirpated in the late 1800s. In 1962 a colony of beavers was found inhabiting Alarka Creek, a few miles southwest of Bryson City, Swain County, North Carolina. Alarka Creek flows into Fontana Reservoir, approximately three miles south of the park boundary. In April 1966 beaver dams were discovered in a small branch of Eagle Creek within the park boundary. On April 7, 1968, a beaver was seen near the mouth of Pinnacle Creek on Eagle Creek; others were observed along the lower reaches of Hazel Creek. Both localities are well within the boundaries of the park. Since then, beavers have been documented in many lower-elevation sites including Abrams Creek, Greenbrier, and Deep Creek. Yet even in the lowlands, most park streams offer only marginal beaver habitat. Because of the area's steep gradients and flash floods, beavers can build dams only on the slowest-flowing side streams. Consequently, nearly all park beavers live in dens dug into stream banks rather than the classic lodges in beaver ponds. Beaver sign, including trees gnawed and felled by beavers, is easily observed along certain streams. Beavers consume the underbark and buds of trees such as yellow-poplar, river birch, black birch, sycamore, and dogwood. The occurrence of beaver in the park is probably the result of introductions made by Tennessee and North Carolina wildlife agencies to rural areas in their states.

White-tailed Deer

Due to persistent hunting, running by dogs, disease, and predators, white-tailed deer almost disappeared from the park by about 1930. In fact, Superintendent Eakin stated in a letter dated July 13, 1931, "From the best information I can gather on the ground there is not one deer left." At that time there was discussion about stocking the new park with white-tailed deer from the Pisgah National Forest. In his book *The Great Smokies: From Natural Habitat to National Park* (University of Tennessee Press, 2000), Daniel Pierce made reference to two letters concerning this discussion in the files of the National Archives and Records Administration in College Park, Maryland. Since there is no evidence in park files or in park correspondence that restocking ever occurred, a request was made in March 2007 to the National Archives for copies of the letters cited by Pierce. Initially, only one could be located; the second reference, as cited by Pierce, was incorrect, but additional searching located not only the second letter but five others relating to this matter

White-tailed deer are primarily active in early morning and early evening throughout the year. They generally spend the daylight hours in concealing cover and bed down in a different site each day. They have no permanent shelter or den.

dating from March 30, 1930, to July 17, 1931. In a letter addressed to the director of the National Park Service dated March 30, 1931, Superintendent Eakin outlined a plan to stock 675 deer (between 25 and 100 deer per year) from 1933 to 1940. However, in a letter dated June 29, 1931, W. C. Henderson, acting chief of the Bureau of Biological Survey, stated, "We do think that the introduction of any more large animals into the area would be unwise." Thus, no evidence has been found that stocking ever took place.

Perhaps only as many as 50 animals could be found in the entire park area in January 1942. However, the formation of the park provided a refuge, and numbers of deer gradually increased, especially in Cades Cove. Deer density in Cades Cove peaked at 43 per km^2 in the late 1970s, declined slowly thereafter, but still remains quite abundant today. Intense browsing by the high densities of deer in Cades Cove has been identified as an important factor in reducing the diversity, richness, and density of woody understory vegetation during a two-decade period (1977–78 to 1995–2002). In addition, the use of deer exclosures in Cades Cove has shown that forests have experienced regeneration failure as a result of the heavy browsing. This will likely have drastic effects on the structure of forest stands in the future.

A companion study investigating the long-term response of spring flora to chronic deer herbivory showed their potential to drastically alter herbaceous woodland plant communities. The recent recovery of plant species as deer densities slowly declined has been largely restricted to species that were able to persist under intense herbivory (mid-1970s to the mid-1980s). However, these species have increased in number within deer exclosures, suggesting continued impacts

White-tailed deer fawns will retain their white spots for about three to four months. The spots are gone by late summer during the first molt to winter pelage.

by deer on the plant community outside the exclosures. Early-flowering, browse-sensitive plants such as trillium and other liliaceous species showed little to no recolonization in the exclosures.

Beginning in 2002, methods to monitor density and health of the deer herd in Cades Cove have included Distance Sampling (use of range finder and compass to determine number of deer per hectare) and herd health checks. Previously, nighttime roadside counts and herd health checks were used to monitor the relative density and health status of the herd. The average number of deer observed during nighttime roadside counts was 85 in 2000, 61 in 2001, and 46 in 2002. The average relative density estimate for 2000 (0.18 deer/hectare) was slightly higher than for 1999 (0.16 deer/hectare). For 2004 it was 0.15 deer/hectare. Deer density estimates are lower than those reported in the mid-1980s, which may be due to higher mortality from coyotes and other predators and changes in the management of the Cades Cove fields. Abomasal parasite counts (APC) are performed in the fall in order to monitor the health of the herd. (The abomasum is one of the four compartments of the stomach of a ruminant mammal such as a deer.) The average parasite count in 2005 was 784, which was lower than 2003 (1,260), indicating a good probability the herd is within the nutritional carrying capacity. An average above 1,500 would indicate a good probability that the herd is exceeding nutritional carrying capacity.

Aversive conditioning (flares, loud noises) is used as a management tool for roadside deer in Cades Cove. Mowing is restricted from June 1 to July 15 in an effort to minimize any impacts that the activity might have on newborn deer.

Rodents

Over one-third (27) of the mammals in the park belong to the family of rodents, or gnawing mammals. One of these, the eastern woodrat, inhabits rock crevices of cliffs and rugged talus slopes. They may also be found in wooded bottomlands, swamps, and in outbuildings or abandoned structures. They have been recorded from only about a dozen localities in the park, all below 2,500 feet. Woodrats habitually collect all sorts of unusual objects and debris which they horde in their nests. While they accumulate all sorts of things including rags, bleached bones and skulls, eyeglasses, and false teeth, they have a particular fondness for shiny objects—nails, spoons, paper clips, coins, bits of foil, broken glass, and the like. When an object is taken, something is often left in its place—a few nuts, an acorn, a pine cone, or a pebble. Because of their habits, these rats are also referred to as packrats, particularly in the western states.

The woodchuck is the largest rodent in the park. It is also one of three rodents that hibernate during the colder winter months. The other two are the meadow jumping mouse and the woodland jumping mouse. Each of these mammals stores fat and does not awaken during hibernation. Chipmunks may become inactive during severe winter weather and, in some regions, may enter into a deep sleep, but they do not store fat and must awaken periodically to feed on stored food.

The golden mouse, which gets its name from its bright golden fur, is one of the most beautiful and unique mice in the park. It occurs in highly localized populations, usually inhabiting wooded areas having an extensive understory of greenbrier or honeysuckle. Unlike most mice, this species is semi-arboreal with its semi-prehensile tail serving as a balancing and stabilizing organ. As the mouse travels along vines and branches, the tail is used for balance. When it pauses, the tail encircles the branch or vine. It builds a spherical nest of leaves, shredded bark,

The woodchuck is also known as the "groundhog" and "whistle-pig." In the park, it has been seen from the lowest elevations to approximately 6,300 feet. Woodchucks are solitary and most active in early morning and late afternoon. They have excellent eyesight and are able to climb trees to escape an enemy.

and grasses that is six to eight inches in diameter and is located 10 to 25 feet above ground in the fork of a tree or in greenbrier, honeysuckle, or wild grapevines against the trunk of a tree. Very little was known about this southeastern species in the 1960s, so I chose it as the subject for my Ph.D. research and dissertation.

My main study area was in the Cosby section of the park, where I spent the summer months as well as each spring and Christmas vacation gathering data on three populations. Other areas that were studied less intensively included Greenbrier, Deep Creek, Smokemont, and Cherokee Orchard. In just one two-year period, I amassed a total of 27,450 trap nights of effort (one trap night equals one trap set for one night). My studies covered nest construction and placement, food habits, population size and population fluctuations, home range, longevity, parasites, reproduction, growth and development, and molting. Several marked mice in the wild population survived longer than a year. In addition, I maintained a colony of breeding adult golden mice in order to obtain reproductive and growth and development data. One of the captive individuals lived for eight years, five months—the longest recorded life span for any native North American murid rodent. My studies on the ecology and life history of this species have resulted in five publications.

Trapping for small mammals can be both exciting and frustrating. Traps are normally baited with either peanut butter or potted meat. Every morning it is exciting to check the traps that have been set overnight. Every time that you come upon a closed trap, it is like opening a Christmas present—you never know what it may contain. Hopefully, it will be a small mammal, but sometimes it may be a millipede, a grasshopper, a snail, or a salamander. And sometimes, nothing at all. If the trigger mechanism is set too fine, the trap may close with the slightest vibration from the wind or rain. Trapping can also be frustrating and expensive if a bear, boar, or raccoon finds the trapline and follows it, tearing open each aluminum trap to get at the bait. In the 1960s near Cosby, a bear destroyed about 20 traps out of one trapline containing 50 traps. In September 2001, while trapping on the Ravensford Tract near Cherokee, we lost approximately 20 traps in one night to one or more wild boars. In August 2007 at Tremont, a black bear destroyed 21 out of 50 traps and damaged 5 others.

Five kinds of tree squirrels are found here. The most frequently encountered squirrel in the high-altitude forests is the red squirrel or "boomer," whereas the most common squirrel at middle and low elevations is the eastern gray squirrel. Fox squirrels are rare. Two species of flying squirrels occur in the park (see additional discussion under Endangered Species—chapter 7). Flying squirrels are incapable of true flight; rather, they glide from a high perch to a lower perch. They are nocturnal and are not seen by most park visitors.

An interesting event occurred on August 30, 1965, along a dirt road in the Deep Creek area while I was searching for golden mouse nests. In an area of pine trees and dense honeysuckle, I located what appeared to be a golden mouse nest on the limb of a pine tree approximately 12 feet above the ground. In my attempt to examine the nest more closely, it was inadvertently dislodged and fell to the ground. Inside were not golden mice but four young southern flying squirrels approximately three weeks old. Each was in its own compartment or depression in the nest. I placed the nest and young squirrels temporarily in an open, empty wooden box that was used for mammal live traps. While I was looking at the young squirrels near my car, the mother appeared and kept advancing to within approximately two feet of me. I backed off, and she climbed inside the box, picked up one young squirrel and began making her way through and along trees to a cavity or crotch between two branches in a tree approximately 40 feet from the original nest where she apparently had a second nest. There was no hesitation on her part. After depositing the first young, she returned for each of the others. After retrieving all four young, she returned, sniffed the ground where the box had been sitting, jumped onto the tire of my car, then under the bumper and fender. Finally, she ran up and down several trees and made her way back to the tree cavity where she had taken her young. During this last trip when she was not carrying any young, she made glides of 25 to 30 feet. After I replaced the original nest, she investigated it one time before returning to her secondary nest. The maternal instinct is strong in most mammals, but to witness an event like this at such close range was extraordinary.

In 1993 the Sin Nombre strain of hantavirus caused a much-publicized cluster of human deaths in the Four Corners region of the southwestern United States. In 1994–1995, rodent populations were surveyed in 39 national parks, including Great Smoky Mountains National Park. The sampling in the park was limited to only three nights of trapping with 50 rodents captured; however, two of these rodents were seropositive, suggesting that they had been exposed to Sin Nombre or a similar strain of hantavirus. In 2000 and 2001, researchers from the Department of Comparative Medicine at the University of Tennessee College of Veterinary Medicine surveyed rodent populations in the park and confirmed the presence of a strain of hantavirus. The strain was named Newfound Gap Virus.

Shrews

Shrews are among the smallest mammals, with an adult pygmy shrew weighing only approximately one-twelfth of an ounce (2.5 grams). Shrews possess long tapering snouts, tiny eyes that are probably capable of only limited vision, and ears that are barely visible. Hearing and smell are acute. Eight species of shrews have been recorded within the park, seven of which are terrestrial and feed on insects, spiders, worms, and other invertebrates. Shrews have such a high metabolism that some studies have shown that they must eat over two times their weight in food every day. The water shrew, however, is semi-aquatic and readily takes to water. It can swim, dive, float, and run along the bottom of a stream, and has even been observed running upon the surface of the water for some distance. It lives beneath

In the park, short-tailed shrews can be found in almost all kinds of habitats at all elevations. They are active mainly at night and have poor eyesight but highly developed senses of smell and touch.

Moles may be active at any hour and during all seasons. They feed primarily on earthworms and other invertebrates.

the overhanging banks and in rock crevices along the edges of swiftly flowing mountain streams. The large hind feet are fringed with bristles, an adaptation that makes them efficient swimming paddles. The dense fur traps tiny air bubbles which gives the little animal a silvery sheen as it swims through the water. The bubbles so increase the shrew's buoyancy that, the moment it stops actively swimming, it pops like a cork back to the surface. They feed primarily on small aquatic organisms—insects, fish eggs, and small fish—that they capture while swimming (see discussion about this species on Andrews Bald, page 94.

The short-tailed shrew is the only poisonous mammal in North America. The poison is produced by the submaxillary salivary gland and is present in the saliva. It acts as a slow poison and immobilizes insects and other prey. Immobilized insects remain alive for three to five days and provide a source of fresh, non-decomposing food. Few records are available concerning the effect of this poison on humans, although the bite may cause considerable discomfort and has been known to produce local swelling.

Moles

Three species of moles occur in the park. These mammals are highly specialized for subterranean life. Their soft, silky dense fur lies equally well when brushed either forward or backward, an adaptation to facilitate movement in either direction in the underground burrow. (This is also true of the fossorial woodland vole, which spends much of its time in underground burrows.) The short front limbs possess feet that are greatly enlarged for digging. The forefeet are at least as broad as they are long, and the palms face outward. The tiny, degenerate eyes are concealed in the fur and are covered by fused eyelids. They can distinguish light from dark but little else. External ears are absent. Star-nosed moles are the rarest of the three species in the park. They are semi-aquatic and prefer low, wet areas such

Bears will not work harder than necessary to secure food. They will locate the easiest available source. Beginning in the late 1960s, the Park Service installed bear-proof garbage cans in an effort to encourage the bears to depend on natural foods.

as wet meadows, marshes, and low, wet ground near streams. They are excellent swimmers, and some of their burrows often have underwater openings.

Black Bear

After a summer visit to Great Smoky Mountains National Park, the chances are that a visitor will remember the bears above all else. Since the park is a wildlife sanctuary, decades of protection served to break down the wild bears' fear of man. Instead of depending on their own resources for a living, many bears patrolled park roads and campgrounds, where they rummaged through garbage cans, tore open picnic baskets and ice chests, and received food (an illegal practice) from some naive and foolhardy park visitors. Whereas an average adult male black bear in the park weighs 240 to 260 pounds, some "garbage" bears over 400 pounds have been recorded. On December 8, 1998, a record-sized bear weighing 620 pounds and measuring 89 inches from nose to tail was killed by a poacher just inside the park boundary along Ski Mountain Road. The bear had a long history of sightings in downtown Gatlinburg, where he had frequented trash cans and dumpsters. In the mid-1960s, I can remember seeing eight to 10 black bears on a single trip along the transmountain road from Sugarlands to Cherokee. And, in most cases, each sighting would be accompanied by a "bear jam" of cars in both directions as visitors were anxious to see, photograph, and sometimes feed the most well-known symbol of the Smokies.

Black bears eat mostly berries, nuts, insects, and animal carrion. They have good color vision and a keen sense of smell. In addition, they are good tree climbers, can swim very well, and can run 30 miles per hour. Although adults can weigh several hundred pounds, newborn cubs tip the scales at just eight ounces.

In the past, numerous visitors were scratched or bitten by bears that were enticed by food to come closer than they should. It is illegal to feed any wildlife in the park.

During the past 15 years, there have been a lot of significant improvements to the bear management program, for example, converting from trash cans to dumpsters, earlier closures in picnic areas, later work schedules for maintenance employees, better literature/signs, more aggressive or proactive management, backcountry food storage cables, and aversive conditioning of bears losing their wild behavior.

In the late 1960s, in order to dissuade bears from seeking food from garbage cans and from visitors, the park service began using bear-proof garbage cans. However, given the number of visitors and the volume of garbage, the cans often were overflowing with garbage. In the early 1990s, the park installed bear-proof dumpsters to secure the high volume of garbage and human food. These containers have proven highly effective and have caused most bears to again subsist primarily on natural foods such as acorns, hickory nuts, beechnuts, and berries. Hikers in the backcountry and along the Appalachian Trail use food storage cables at the shelters and campsites to suspend their food in such a way that it is unavailable to bears.

In June 2000 an ordinance mandating animal-resistant containers in certain areas of the City of Gatlinburg (i.e., areas adjacent to the park) was implemented to address the problems of bears and garbage in areas outside the park. In October 2000 the Tennessee Wildlife Resources Commission passed a proclamation, which also was aimed at reducing nuisance bear problems around the Gatlinburg area. The proclamation made it unlawful to feed black bears or leave food or garbage in a manner that attracts bears as well as any other indirect or incidental feeding of bears. The proclamation was the result of a joint effort by the committee on nuisance bears consisting of personnel from the City of Gatlinburg, the University of Tennessee, the Tennessee Wildlife Resources Agency, the National Park Service, and private citizens.

In 2006 the University of Tennessee completed its 38th year of studying the park's black bear population. This study, which was led by Dr. Mike Pelton for many years, represents the longest continuous study of this species. Some of the information that follows has been gathered during this study.

As the weather cools in autumn, bears begin eating considerably more food, especially acorns, to provide nutrition for their winter sleep. In the park, black bears become "dormant" during the colder winter months. Their heartbeat slows from 40 to 50 beats per minute to eight to 10. Their body temperature drops from 102 degrees to about 95. However, their overall metabolism remains fairly high. This is not "hibernation" in the technical sense. The body temperature of true hibernators such as woodchucks, jumping mice, and some bats falls much lower. If bears truly hibernated, females would be unable to give birth in January and nurse their offspring for several months until they left their dens in April or even May. Den sites include inside large, standing, hollow trees; beneath or inside logs; in caves; and other sheltered areas. Bears do not eat, drink, urinate, or defecate during this period of dormancy. A tough, fibrous fecal plug forms in the anal region of the bear's lower digestive tract and remains until it is expelled after the bear leaves its den in the spring.

During years when the fall hard mast (acorns, nuts) crop is poor, the impact on bears occurs during the following year. Bears, particularly adult females with yearlings, emerge from their dens in relatively poor condition. They move extensively during the summer months in search of food. The summer of 2000 was such a time. University of Tennessee researchers conservatively estimated the black bear population at 1,796 individuals, which was slightly less than reported in 1999. The 2001 population estimate was approximately 2,000 bears; the 2002 estimate was approximately 1,700 bears; the 2003 estimate was approximately 1,350; the 2004 estimate was approximately 1,625; the 2005 estimate was approximately 1,250; and the 2006 and 2007 estimates were approximately 1,500 bears each. At almost two bears per square mile, the park's bear population is the densest in North America.

Visitors occasionally are injured by black bears. Most injuries result in scratches elicited by attempting to feed a bear and/or by attempting to invade its space to take a photograph. Feeding bears is misguided kindness; the bears come to expect such generosity from everyone and, consequently, trouble is imminent. Park regulations prohibit the feeding of bears and is punishable by fines and/or arrest. Whereas wild bears have a life expectancy of 12 to 15 years, "panhandler" bears live only about half as long.

Nonlethal aversive conditioning techniques such as trapping and releasing on site and the use of pyrotechnics and rubber/plastic projectiles have been used as management tools for some nuisance bears since 1991, conditioning them to avoid people. Wildlife managers capture the bear at night and induce sleep. They tag its ear, pull a small, nonfunctional tooth, take a blood sample, tattoo its mouth, and take body measurements. The bear is then released in the area where it was caught. When the bear returns, it remembers that scavenging for garbage was a bad experience.

The first and only fatality from a black bear in the history of the park occurred on May 21, 2000. A 50-year-old schoolteacher from Cosby was killed and partially consumed by a black bear approximately 3.5 miles up the Little River Trail from its trailhead at Elkmont. Park law enforcement rangers shot and killed two bears found at the scene, an adult female (111 lbs.) and a female yearling (40 lbs.). The bears had little body fat, but no other known abnormalities.

Park wildlife management biologists are dedicated to ensuring the well-being of wildlife in the park. I have known and worked with them for many years and can attest to their professionalism. For example, in April 2005, a two-year-old male bear was released back into the park in Sugarlands after nine months of rehabilitation at the Appalachian Bear Center (ABC) in Townsend, Tennessee. The bear was originally captured on July 8, 2004, near park headquarters, weighing only 28 pounds; it also had a broken femur. The bear was taken to the University of Tennessee College of Veterinary Medicine, and surgery was performed to install a plate to stabilize the femur. The bear gained 161 pounds during its stay at ABC. The bear was ear-tagged to enable biologists to track its movements. This bear *did* go over the mountain—and not just to see what he could see, either! In the fall, he showed up at the Swag Resort in Waynesville, North Carolina, shaking ripe apples from a tree. His extensive travels show that he is thriving.

Many black bears possess remarkable homing abilities. They are often able to return home after being transported many miles from their original capture

Bear #75. Photo by Ken Jenkins.

Animals ~ 157

Table 6.3. Movements of relocated black bear #75, July 1988–July 1990.

Date	Capture Location	Relocation Area
July 30, 1988	Cades Cove Mill, TN	Cataloochee, GSMNP
August 9, 1988	Park Headquarters, TN	Ocoee Wildlife Mgt. Area, SE TN
September 1, 1988	Cades Cove, TN	Cataloochee, GSMNP
June 23, 1989	Cades Cove, TN	Ocoee Wildlife Mgt. Area
July 22, 1989	Cades Cove, TN	Sullivan County, NE TN
August 4, 1989	Johnson City, TN	Carter County, NE TN
May 11, 1990	Cades Cove, TN	Geo. Washington Nat. Forest, VA, approx. 400 miles from Cades Cove
June 11, 1990	Pearisburg, VA	Geo. Washington Nat. Forest, VA
mid-June 1990	Roanoke, VA	Mountains nearby
July 18, 1990	Johnson City, TN	Carter County, TN
July 21, 1990	Unicoi County, TN	Buried; had been shot by poachers.

point. The journeys of one such adult bear, identified as Bear #75, over a period of 24 months were remarkable (Table 6.3). Bear #75 has the most extensive relocation history of any panhandler black bear handled by park personnel. This was an above-average-size bear that at one point in his travels weighed 367 pounds. According to Kim DeLozier, park wildlife biologist, Bear #75 traveled in excess of 1,500 air miles in an effort to return to its home territory in the park. But neither bears nor humans travel in straight lines in the rugged hills of the southern Appalachians, so his actual mileage was far greater.

Another bear with a unique history is Bear #287, known as "the Troublesome Cub." The following unique history of this 17-year-old female was provided by Bill Stiver, park wildlife management biologist:

> Bear 287 was originally captured and released on site in Cades Cove picnic area on July 3, 1997. During the first capture, bear 287 was fitted with a radio-collar as part of a research project evaluating the effectiveness of capture and release as a form of aversive conditioning for nuisance bears. On the morning of July 22, 1997, bear 287 managed to get inside a broken dumpster in Cades Cove picnic area and was accidentally dumped into a garbage truck. The truck went to several other locations in the park and the bear was compacted multiple times as trash was emptied into the truck. Later that afternoon, bear 287 was dumped, along with a heap of trash, inside the Sevier County compost facility. The bear climbed the walls inside the building and was later tranquilized by an officer from the Tennessee Wildlife Resources Agency (TWRA). Once immobilized, bear 287 got hung in the rafters and the TWRA officer and workers from the compost facility had to use a bucket truck to remove the bear from the rafters. The TWRA officer notified the park about bear 287 and she was transported to a holding facility at Park headquarters for

observation. Bear 287 was very lethargic for two days and on July 24, 1997, was taken to the University of Tennessee, College of Veterinary Medicine (UTKCVM) for examination. Veterinarians determined bear 287 had a tear in her urinary tract and they removed urine from her abdomen and around her spleen. They also indicated that the animal had a lot of bruising. The prognosis didn't seem good and it was decided to take bear 287 back to Park Headquarters for observation for a few more days. Bear 287's behavior improved and on July 29, 1997, bear 287 was taken back to the UTKCVM for reexamination and possible surgery to repair the urinary tract. However, the tear healed and the surgery was not necessary. On July 30, 1997, bear 287 was released at Mount Sterling Gap, approximately 40 miles from the dumpsters in Cades Cove. Bear 287, "The Troublesome Cub," was not seen again until May 17, 2004, when she was seen in a tree at the entrance to Cades Cove picnic area. Wildlife staff had to use cracker shells and bean bag rounds to get her away from the area. For the next two weeks she was observed in the picnic area during the evening and was even seen getting into another broken dumpster. Bear 287 was recaptured on June 1, 2004, and was again taken to the UTKCVM, this time for broken teeth. One loose tooth and one broken and decayed tooth were extracted. On June 2, 2004, bear 287 was relocated to the Cherokee National Forest, about 60 miles from the dumpsters in Cades Cove. On August 13, 2005, Bear 287 was captured again in Cades Cove picnic area. Once again she was getting into garbage. However, this time she was accompanied by a 20 pound male cub. Bear 287 weighed 161 pounds, which is rather large for an adult female during summer. Bear 287 and her cub were moved to the Cherokee National Forest.

They were not seen during 2006 or 2007. A popular children's book entitled *The Troublesome Cub in the Great Smoky Mountains* was written in 2001 and recounts the adventures of this bear up to that time. The book is available in the park's bookstores.

The ears of Rafinesque's big-eared bat are over one inch long and are joined at their base. Like all species that occur in the Park, these bats feed on flying insects. They have been known to reach at least 10 years of age.

Bats

The 11 species of bats constitute 15 percent of all mammals recorded in the park. All feed exclusively on insects. During the colder months when flying insects are unavailable, bats must either hibernate or migrate to warmer areas. Eight species found in the park are known to hibernate. One shaft of an abandoned copper mine site in the southwest quadrant of the park is the site of a maternity and hibernating colony for Rafinesque's big-eared bat—the largest known colony in the world. Only three of the park's bats—the red bat, hoary bat, and silver-haired bat—are migratory.

Bats are the only mammals capable of flying. Flight membranes, which are actually extensions of the skin of the back and belly, connect the body with the wings, legs, and tail. The membrane extending from the tail to the hind legs is known as the interfemoral membrane. Unlike birds, bats use both wings and legs during flight. Other modifications for flight include greatly elongated fingers to provide support for the wing membrane, a keeled sternum for the attachment of the enlarged flight muscles, and fusion of some vertebrae.

Bats use echolocation, a system somewhat similar to radar. They emit ultrasonic calls, far above the range of human hearing, that are reflected from objects ahead of them. Their very sensitive hearing apparatus instantly analyses the echo, giving the bat the size and distance of the object. Thus, bats are able to avoid obstacles and find food in total darkness. Different species can be distinguished by differences in the structure of their echolocation calls.

On August 12, 2004, an Iowa woman became the only known person ever to be bitten by a rabid animal in the park. The woman was hiking on the Old Sugarlands Nature Trail when she was bitten by a bat (eastern pipistrelle), which later tested positive for rabies. According to reports, the woman was hiking with a group of a dozen people when the bat began flying erratically around her and eventually landed on her fanny pack. She apparently contacted the animal with her elbow and received a small puncture wound on her left elbow. Other members of the group were able to catch and kill the bat and bring the carcass to the Sugarlands Visitor Center. Park biologists took the bat to the Sevier County Health Department for testing, which was positive for rabies. The woman underwent a series of rabies shots as a precaution. The park has been testing certain bats that have turned up dead or that were acting suspiciously in high human-use areas, but this is only the fifth one to come back positive for rabies. The first (species unknown) was in June 1989 at the Oconaluftee Visitor Center in North Carolina, and the second was a red bat (*Lasiurus borealis*) found on the road in front of the Little River Ranger Station in Sugarlands on May 10, 2003. Sick bats were collected in July 2006 at the Oconaluftee Visitor Center (big brown bat, *Eptesicus fuscus*), and in September 2006 at park headquarters (eastern small-footed myotis, *Myotis leibii*). Both bats tested positive for rabies.

For in-depth discussions of the natural history and biology of each species, go to discoverlifeinamerica.org/atbi/species/animals/vertebrates/mammals.

CHAPTER 7

ENDANGERED SPECIES

There is nothing in which the birds differ more from man than the way in which they can build and yet leave a landscape as it was before.

—Robert Lynd (1923)

There are currently three species of plants, one invertebrate, and 11 species of vertebrates in the park that are listed as endangered or threatened by the U.S. Department of the Interior (Table 7.1). One additional species, the peregrine falcon, has been delisted but is still being monitored. Many other species are classified by the states of Tennessee and North Carolina as endangered or threatened on a local level.

Rock Gnome Lichen

The rock gnome lichen occurs only at high elevations on boulders in streams or on other rock surfaces located in high-moisture environments. In an effort to determine whether acid precipitation was having any effect on the few small colonies that exist in the park, colony sizes were measured in 2001 and compared to their size in 1994. Six of eight patches increased in percent cover (range 6 percent to 13 percent), and one colony remained stable. Only one colony decreased in size, and the decrease was minimal (3 percent). Due to the small sample size, these results should be interpreted with caution.

Spreading Avens

The park has only one population of this globally rare species, which grows at high elevations on rocky outcrops or openings. A 1989 census revealed about

Do You Know:
How many endangered and threatened species occur in the park?
Where Indiana bats occur in the park?
Where northern flying squirrels have been found in the park?
Whether a breeding population of cougars exists in the park?

Table 7.1. Federally endangered and threatened plant and animal species in the park (E = endangered; T = threatened; DM = delisted, monitored; Ex = extirpated)

Species	Common Name	Status
Plants		
Gymnoderma lineare	Rock Gnome Lichen	E
Geum radiatum	Spreading Avens	E
Spiraea virginiana	Virginia Spiraea	T
Animals		
Microhexura montivaga	Spruce-Fir Moss Spider	E
Cyprinella monacha	Spotfin Chub	T
Noturus baileyi	Smoky Madtom	E
Noturus flavipinnis	Yellowfin Madtom	T, E
Etheostoma percnurum	Duskytail Darter	E
Haliaeetus leucocephalus	Bald Eagle	T
Falco peregrinus	Peregrine Falcon	DM
Picoides borealis	Red-cockaded Woodpecker	E
Myotis sodalis	Indiana or Social Myotis (Bat)	E
Glaucomys sabrinus coloratus	Carolina Northern Flying Squirrel	E
Canis rufus	Red Wolf	E, Ex
Canis lupus	Gray Wolf	E, Ex
Puma concolor couguar	Eastern Cougar	E, Ex

five large patches growing on vertical cliff ledges. The exact count was difficult as the plants are rhizomatous, and ropes are needed to survey the entire area. At one time, the population was estimated to consist of about 25 to 50 plants. As of the summer of 2005, the population appeared stable with many flowering stalks.

Virginia Spiraea

This species is found in just one location in the western portion of the park. It tends to like cobbly scoured areas right on the riverbank. As of 1992 the park had four clones spread out along a several-mile stretch of river. They have flowered in the past but rarely set fruit. It is thought that the clones (several meters long and wide) are pieces of the same or few individuals. A survey in August 2006 revealed that one colony (27 m by 7 m) was still intact, but two smaller colonies were no longer present. All remaining plants looked very healthy and were in fruit. One additional colony located farther downstream will be monitored at a later date. At this time, the population has been declared stable.

Spruce-Fir Moss Spider

Microhexura montivaga was first discovered in 1923 on Mount Mitchell. It was first collected on Mount LeConte in 1926 and on Clingmans Dome and Mount Collins in the 1970s. This tiny, tarantula-type spider is light brown to reddish-brown with the largest adults being barely 0.15 inches long. It belongs to a small relict genus of primitive tube-web spiders. Its primary microhabitat consists of humid/moist, but well-drained bryophyte mats growing on sheltered, well-shaded, north-facing rock outcrops and boulders in Fraser fir and/or fir-dominated spruce-fir forest above 6,100 feet.

In the 1980s this species suffered a marked decline on Clingmans Dome, apparently due to the desiccation of its primary microhabitat. This microhabitat change was caused largely by the widespread mortality of mature Fraser fir trees and the resultant opening of the forest canopy.

In 1995 the spruce-fir moss spider was listed as an endangered species under the Endangered Species Act of 1973. Its critical habitat was designated in 2001.

Funding was obtained in 2004 to conduct a status survey. The last limited survey had been conducted in 1991, but because of the dramatic changes in the fir forests, another survey was necessary to determine the location and relative abundance of populations in the park. Sixty sites were searched by Dr. Fred Coyle during the 2004 survey, and *Microhexura* was found at 12 of these sites. Eight of these sites were new locations for this species in the park, thus increasing the number of known localities within GSMNP from seven to 15. Sizable viable populations were found to exist on Mount LeConte, Mount Buckley, and Mount Love. A locality on Mount Chapman is probably the easternmost edge of its range in the park. Population densities were not determined during this survey. At all of these sites, the canopy was relatively open because of the loss of at least some fir and often because steep slope with extensive outcrops prevent high tree density. Spruce-fir moss spider populations can survive within some of these stands even though the trees are

Spruce-fir moss spider. Photo by Frederick A. Coyle.

diseased, presumably because the north-facing outcrops provide their own shelter from the drying effects of the sun.

Spotfin Chub

(See discussion under Reintroductions, chapter 8.)

Smoky Madtom

(See discussion under Reintroductions, chapter 8.)

Yellowfin Madtom

(See discussion under Reintroductions, chapter 8.)

Duskytail Darter

(See discussion under Reintroductions, chapter 8.)

Peregrine Falcon

(See discussion under Reintroductions, chapter 8.)

Red-cockaded Woodpecker

In Tennessee, this species occurs in disjunct relict populations occupying pine or oak-pine associations in predominately upland hardwood forests. It was first reported in the area presently comprising the park by Fleetwood in 1936, who observed birds at three locations in the southwestern region in 1935. All other reports have come from the same region but have been widely scattered over the years.

Red-cockaded woodpecker. Photo by Ann and Rob Simpson.

During a study between May and July 1979, one clan was located on Skunk Ridge in the southwestern portion of the park. Three cavity trees were observed, but only one was active. The active cavity was in a Virginia pine, whereas the two inactive cavities were in a Virginia pine and a shortleaf pine. Suggested courses of action included continued searching for additional colonies, intensively managing the existing colony site by mechanical removal of hardwood species, and immediately suppressing fires that threaten colony sites.

Indiana Bat

Three park caves (Blowhole, Bull, and Scott) serve as hibernacula for Indiana bats, an endangered species. The U.S. Fish and Wildlife Service lists White Oak Sinks Blowhole Cave as a critical habitat for the largest-known hibernating colony of Indiana bats in Tennessee and as one of the largest-known hibernating colonies in the United States. A winter survey of Blowhole Cave in 1999 resulted in a count of 3,084 bats. In 2001 a winter survey of Blowhole revealed 4,548 bats, with 553 bats in Bull Cave and 102 in Scott Cave. In January 2003 a survey of Blowhole Cave revealed 5,564 bats. In February 2005 there were 7,681 Indiana bats in the cave. Three banded Indiana bats were observed in the cave in 2005: a male captured at the entrance of the cave in July 2004; a lactating female captured in the west end of the park on July 5, 2000; and a young-of-the-year female captured west of the park in Blount County, Tennessee, on July 14, 2000.

Indiana bats have been captured in the park during the summer for three consecutive years. During the summer of 2000, Dr. Michael Harvey and researchers from Tennessee Technological University (TTU) found a maternity colony of 23 Indiana bats in a snag in the Forge Creek Road area; this was the first record of a maternity (breeding) colony of Indiana bats in the state of Tennessee and the second in the entire Southeast. (In 1999, TTU researchers had located a maternity colony on the Nantahala National Forest in North Carolina.) In 2001, a second maternity colony of 56 Indiana bats was found in the park in a pine snag near Dalton Gap, and in 2002, another maternity colony was located in a pine snag just east of U.S. Highway 129 near the Tennessee/North Carolina state line. These findings have significantly extended the southernmost range for an Indiana bat maternity colony. Previously, the southern range for maternity colonies extended into Kentucky and southern Indiana. The findings also revealed that some Indiana bats are nonmigratory and reside in the park during the summer.

Northern Flying Squirrel

Under a contract with the U.S. Fish and Wildlife Service, my students and I erected 490 flying squirrel nest boxes in 35 carefully selected high-elevation areas within the potential range of this species in Maryland, West Virginia, Virginia, Tennessee, and North Carolina during the period 1980–1983. A total of 41 nest boxes

Except for being larger, the northern flying squirrel appears similar to the southern flying squirrel. It is nocturnal and feeds on lichens, mushrooms, seeds, buds, fruit, conifer cones, meat, and arthropods.

were erected within the park on Balsam Mountain, near Newfound Gap, and at various sites along Clingmans Dome Road. Prior to this study, only two specimens were known from the park (Blanket Mountain, 4000 ft. elevation, February 10, 1935; West Prong of the Little Pigeon River near Walker Prong, August 22, 1959). Although we located squirrels on Mount Mitchell in North Carolina (first records in over 30 years) and at a new site in West Virginia, we failed to secure any specimens from the park. This study was used as the primary basis for proposing endangered status for this subspecies (Federal Register, 1984). Beginning in 1987 Peter Weigl of Wake Forest University undertook studies of known and

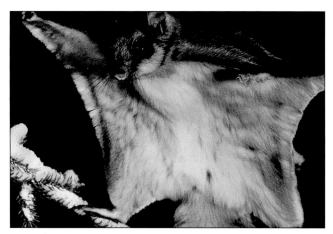

When the flying squirrel launches itself, it spreads its legs, thus drawing taut the folded layer of loose skin along each side of the body. Cartilaginous spurlike supports at the wrists make it possible for the animal to extend the skin fold beyond the outstretched legs. They can glide (not fly) for 100 feet or more.

likely flying squirrel habitat in the park as well as along the Blue Ridge Parkway in North Carolina. These studies resulted in locating northern flying squirrels at two of the three known sites in the park as well as at 19 new sites. Considerable data was also obtained concerning sex ratios, body size, body weight, longevity, reproduction, social behavior, food habits, home range, nest use, parasites, and habitat characteristics.

In 1997 personnel from the North Carolina Wildlife Resources Commission and the park initiated a monitoring program to further determine the distribution and status of northern flying squirrels in the park. Nest boxes were erected in seven high-elevation areas. Initial monitoring in 1998 revealed no squirrels. However, monitoring efforts during January 1999 revealed squirrels in three areas including Balsam Mountain Road, Mount Sterling, and the Appalachian Trail near its intersection with Sweatheifer Trail. Monitoring efforts during 2000 detected squirrels in two of the same areas, Balsam Mountain Road (two females, one male) and the Appalachian Trail near the intersection of Sweatheifer Trail (one female, one male). These findings suggested that the distribution of this species in the park was much greater than originally thought. Monitoring efforts in 2001 resulted in no squirrel captures. In 2002, squirrels were captured at Indian Gap (n=1) and Sweatheifer Trail (n=6). In 2003, eight squirrels were captured in three locations; five were recaptures. In 2004, 14 squirrels were captured in three locations; two were recaptures. In 2005, there were three captures at one location. No squirrels were found in 2006.

Red Wolf

(See discussion under Reintroductions, chapter 8.)

Eastern Cougar

Mountain lions, also known as panthers, painters, pumas, and cougars, once had the widest range of any mammal in the Western Hemisphere and were found throughout the area now encompassed by the park. The last cougar killed in the Great Smokies was in 1920. As the story goes, Tom Sparks was attacked by a cougar while herding sheep on Spence Field. He inflicted a deep knife wound in the cougar's left shoulder. Several months later, a cougar was killed near what is now Fontana Village, and its left shoulder blade had been cut in two. It was believed to be the same cat that Sparks had wounded.

Culbertson (1977) examined the status and history of this species in the park. Twelve sightings were reported for the years 1908–1965 and 31 sightings for the years 1966–1976. Culbertson stated:

> The number of lion sightings through the years suggest that the mountain lion may never have actually been extinct in the Great Smoky Mountains area. The lion

Eastern cougars (mountain lions) are solitary, mainly nocturnal animals but may be active at any hour and at any season. Deer are the preferred food.

may have been able to maintain itself in small numbers in the more inaccessible mountainous regions in or around the park. The present lion population could be derived in part from this small reservoir. . . . It is believed that there were three to six mountain lions living in the park in 1975, and other lions were reported to the southeast and northeast of the park as well. Lions were seen most frequently near areas of high deer density.

Park files contain many interesting reports of purported mountain lion sightings. Every year, several additional reports are received. If, in fact, the animals being observed are mountain lions, they may be part of the original population as Culbertson suggested, or more likely, they may be captive animals that have either escaped or been released.

Since 1998 I have been working intensively in an attempt to ascertain whether a viable population exists within the park. I plot all reliable reports on a map of the park with color-coded pins in order to see if there is any clustering of reports. What do I consider a reliable report? The reported sighting in June 2002 of a mountain lion along the transmountain road by a veterinarian who treats mountain lions in his practice; a sighting in May 2002 by a wildlife photographer working with the All Taxa Biodiversity Inventory and a sighting in the same area by a park visitor approximately two weeks later; an adult cougar lying on a hillside for approximately five minutes that was seen and photographed by 15 to 20 park visitors in August 2006. We have erected and maintained rubbing pads from Davenport Gap to Cades Cove in an attempt to secure hair for DNA analysis. The pads have proven successful in securing hair samples from a wide variety of mammals including bears, gray fox, red fox, coyotes, wild hogs, and others, but not mountain lions. I have also employed remote heat-sensing infrared cameras in areas of reliable sight-

Often mistaken for a mountain lion, the bobcat has a short, broad face with prominent, pointed ears. The tail is short with several blackish bars on the dorsal surface just in front of the tip.

ings. We have secured good photographs of mammals including deer, gray squirrels, raccoons, opossums, skunks, turkeys, and coyotes but no mountain lions. If we could capture a photograph of a young cougar, it would be a start to proving that the park contains a reproducing population. Kittens have been reported in the park on only two occasions—both in the 1970s. Small tracks accompanying large tracks were observed in the Cataloochee area. A female with two kittens was observed by a large group of visitors at the old Chimneys Campground (now the Chimneys picnic area).

A third technique involves searching for scats (feces) of cougars. I am always on the alert for scats as I hike. However, I can only find the ones that I can see. Therefore, I have trained my part-Labrador rescue dog, Brandi, to use her keen sense of smell to help locate scats. She is amazing. She locates carnivore scats that I would never see—ones that are buried under leaves and that are as far as 10–15 feet off the trail. The majority of scats that we have located have been identified through DNA analyses as being from coyotes and/or bobcats. No cougar scats yet, but there may be one around the next bend in the trail.

I now possess two photographs of mountain lions taken by visitors in the park—one from Greenbrier and one from Cades Cove. The Cades Cove photograph shows a side view of the entire animal. Thus, we know there are mountain lions in the park. However, we still do not know where they might have come from nor do we have proof of a breeding population.

In January 2007 the U.S. Fish and Wildlife Service initiated a review of the status of the cougar in the eastern United States. Currently, the official federal status of the eastern cougar is that it is extinct. The last such review was in 1982.

CHAPTER 8

REINTRODUCTIONS

Every creature is better alive than dead, men and moose and pine trees, and he who understands it aright will rather preserve its life than destroy it.

—Henry David Thoreau (1848)

One of the mandates of the National Park Service is that park officials are to protect and preserve native species. When a species is eradicated by human-caused activities, officials are to restore that species if it is feasible.

Reintroduction programs are a high-risk conservation strategy for restoring populations and biodiversity. The success of these programs often depends on the ability to identify suitable habitat within the species' former range. Reintroduction science is a newly evolving field that deals with habitat loss and degradation and the consequences of climate change, all of which are likely to reduce the survival of many species.

For reintroductions to be successful, many factors must be considered and evaluated. Long-term assessment is vital, both pre- and post-release. Habitat quality, food availability, competition, predation, and possible diseases are all factors that must be thoroughly evaluated by means of sound scientific research. Education and cooperation between interested parties can be the most important step in creating a successful reintroduction plan, especially when dealing with large carnivores and ungulates. Monitoring goals must be set and monitoring techniques must be established for evaluating progress toward a desired outcome. Some reintroductions are successful; others are not.

In the park, reintroduction programs have targeted nine vertebrate species—four fish (smoky madtom, yellowfin madtom, spotted chub, and duskytail

> **Do You Know:**
> Why smoky madtoms were extirpated from the park?
> Why river otter were extirpated from the park?
> When elk were reintroduced into the park?
> Why the experimental red wolf reintroduction was terminated?

darter), two birds (peregrine falcon and barn owl), and three mammals (river otter, elk, and red wolf). Of these nine programs, five have been successful, two have had questionable results to date, and two have been unsuccessful.

Smoky Madtom and Yellowfin Madtom

Madtoms are members of the North American catfish family. They differ from other catfishes, such as channel, blue, and flathead catfishes, and bullheads in having the adipose fin (small fleshy fin between the dorsal and caudal fins) continuous with the caudal fin rather than discreet. Madtoms possess venomous spines and can deliver a painful sting. They are very secretive and have no human food value. Both madtoms are very small but differ in color and size. The yellowfin attains a length of three inches and is beautifully marked with four dark dorsal saddles: one below the dorsal fin origin, a second between the dorsal and adipose fin, a third below the adipose fin which extends onto this fin, and a fourth on the end of the caudal peduncle. The smoky madtom is slightly smaller and has small saddles on top of the head, beneath the dorsal fin origin, between the dorsal and adipose fins, and beneath the adipose fin; the latter saddle extends slightly onto the adipose fin. The ground color of the head and body is light brownish.

In 1957 the dam on the Little Tennessee River that would form Chilhowee Lake on the southern boundary of the park was nearing completion. In order to create habitat for rainbow trout, a non-native game fish that would be stocked in Chilhowee Lake, the Tennessee Wildlife Resources Agency, Tennessee Valley Authority (TVA), National Park Service, and the U.S. Fish and Wildlife Service decided to eradicate nongame, or "trash," fish in "impounded areas." They wanted to restore the park "to a pristine trout state," according to Steve Moore, park fishery biologist.

Abrams Creek is a tributary of the Little Tennessee River. The plan was to rid lower Abrams Creek of carp, an introduced fish from Asia. Since this portion of the

Smoky madtom. Madtoms are primarily nocturnal, feeding mainly on aquatic invertebrates and small fishes. They are difficult to collect, often burying themselves beneath several inches of gravel during the daylight hours. Photo by Conservation Fisheries, Inc.

Abrams Falls in the western portion of the park.

river was good for trout fishing, biologists believed that the trout would do better if the carp in the river were killed. Agents from the wildlife agency put rotenone, a fish toxin, into the waters from Abrams Falls to Chilhowee Reservoir, a distance of approximately 14 miles. Prior to the reclamation project and impoundment, 67 fish species were known to inhabit Abrams Creek; since that time, fewer than 40 fish species have been recorded in Abrams Creek. Most of the extirpated species were small fish that were of little concern to fishermen. Two of those lost were the smoky and yellowfin madtoms. The smoky madtom was originally only known from Abrams Creek, and because of the 1957 project, was presumed extinct when it was formally described in 1969. The yellowfin madtom was historically more widespread in the upper Tennessee River drainage, but was also presumed extinct at the time of its formal scientific description in 1969. Thus, unbeknownst to anyone, the world's only population of the small, secretive smoky madtom was eliminated from the park before it was even described.

Biologists in the process of reintroducing the smoky madtom to Abrams Creek.

From 1957 to 1988 there were no living smoky or yellowfin madtoms in the park. Then, in 1980, smoky madtoms were discovered in Citico Creek in the Cherokee National Forest near Tellico Plains. While studying the biology of Citico Creek smoky madtoms, yellowfin madtoms were found in the same creek in 1981. Citico Creek enters the Little Tennessee River below Chilhowee Dam, on the opposite side of the river from Abrams Creek. The smoky madtom was immediately placed on the federal list of endangered species, and a recovery plan was prepared. Plans were also made to reintroduce two other species that were wiped out in the poisoning of Abrams Creek—the yellowfin madtom and the spotfin chub. Both of these species are listed as "threatened" under the Endangered Species Act. Since the Endangered Species Act of 1973 mandates that agencies formulate restoration plans for endangered species, its reintroduction was part of the Great Smoky Mountains National Park's attempt to meet the National Park Service's mandate stating that park officials are to protect and preserve native species.

In 1982, the Tennessee Wildlife Resources Agency, the National Park Service, the U.S. Fish and Wildlife Service, and the U.S. Forest Service formulated plans to restore the species. All of the madtom work (captive rearing, habitat assessment, stocking, and assessment surveys) was carried out by J. R. Shute, Peggy Shute, and Patrick Rakes of the nonprofit fish research organization Conservation Fisheries Inc. Abrams Creek was assessed to see if it contained elements of preferred habitat for the madtoms. The survey of the creek was favorable. On June 5, 1986, the first egg masses were collected from Citico Creek and transported to a University of Tennessee lab where they were reared. Initially, 200 to 350 eggs from each species were brought to the lab, raised in aquaria, and fed until they were a stockable size.

On August 30, 1988, 155 yellowfin and 118 smoky madtoms were released into Abrams Creek. On September 28, 1989, an additional 174 smoky and 90 yellowfin madtoms were released at two sites in Abrams Creek. An additional 34 smoky and six yellowfin madtoms were retained for captive breeding. In 1990, 151 smoky madtoms were released at three sites in Abrams Creek. No yellowfin madtoms were released in Abrams Creek, but 74 were released in Citico Creek to supplement the existing population. In 1991, 134 smoky madtoms were released in Abrams Creek.

In 1991, two madtoms, one a very large gravid female found under a small slab rock, were found in Abrams Creek. This was the second consecutive year that individuals were recorded in the park.

As of December 2006 a total of 3,239 smoky madtoms and 1,638 yellowfin madtoms had been released into Abrams Creek (Rakes, personal communication). Populations appear to have been successfully reestablished with smoky madtom abundance being nearly comparable to the native population in Citico Creek. The Recovery Plan calls for populations of all four reintroduced species (smoky madtom, yellowfin madtom, spotfin chub, and duskytail darter) to be self-sustaining and relatively stable over 10 years before completion of the project. Significant stockings have ceased with researchers now monitoring population expansion and dispersal.

The poisoning of the river, the impoundment of the lake, the entry of new fish species, and the restocking of four species resulted in a net loss of 17 native fish species from the park, which now has 79 native species and eight introduced species. No other such drastic loss of species is known in the history of the park.

Spotfin Chub

This sleek chub with small eyes and a long pointed snout reaches a length of about four inches. The color of the dorsum ranges from olivaceous to steel blue; a characteristic dark spot is usually conspicuous in the posterior portion of the dorsal fin. Breeding males are brilliant metallic blue with breeding tubercles on the top of their head.

On October 26, 1988, the park service reintroduced this small minnow into lower Abrams Creek. Personnel from the U.S. Fish and Wildlife Service, North Carolina Wildlife Resources Commission, the University of Tennessee, and the National Park Service collected 250 fish from the Little Tennessee River in Swain County, North Carolina, just upstream from Fontana Reservoir. They were transported to Abrams Creek and released.

As of December 2006 a total of 11,367 spotfin chubs had been released in Abrams Creek (Rakes, personal communication). Researchers are fairly confident that this species did not become established probably because Abrams Creek is too small and perhaps too cool above the impounded reach (Rakes, personal communication).

Duskytail Darter

The duskytail darter was first recognized as a distinct species in 1971, but was not scientifically described until 1994. The two-and-a-half-inch fish has a straw to olivaceous colored body, the top of the head is medium to dark gray, and the belly

The duskytail darter is a small, bottom-dwelling fish with a flat head. It reaches a length of approximately 2.5 inches and has 10 to 15 dark vertical bars on the sides of its body. Photo by Conservation Fisheries, Inc.

is dingy white to pale gray. There are 10 to 15 long dark vertical bars on the side of the body. Brilliant gold, fleshy knobs develop on the tips of the dorsal fin spines of males during the breeding season.

The Abrams Creek duskytail darter population was presumably extirpated in 1957 during the reclamation project that eliminated most native species. The U.S. Fish and Wildlife Service, in coordination with the Tennessee Wildlife Resources Agency, National Park Service, U.S. Forest Service, and Conservation Fisheries Inc. has reintroduced this species into Abrams Creek. As of December 2006 a total of 3,430 duskytail darters had been released; the reintroduction appears successful based on an increasing range in Abrams Creek and natural reproduction.

River Otter

Northern river otters, or "orters" as they were known to the mountain hunters, were extirpated in the Smokies due to overharvesting and logging in the early part of the last century. They were last seen in the park in 1927, when three individuals were sighted in the Cataloochee area, although Willis King, an associate wildlife technician, stated in 1937 that otters "occurred in the Smokies less than 10 years ago." Although the reintroduction of the northern river otter to the park had been given serious consideration in the 1960s, it was not until 1986 that a reintroduction program was begun. Between February 26 and March 31, 1986, eleven river otters were obtained from North Carolina, implanted with radio transmitters, and released in Abrams Creek. From December 1988 to March 1990, 14 river otters from South Carolina and Louisiana were released in Little River. Three otters of this latter group crossed the mountains and established home ranges on the North Carolina side of the park. From August 1990 to February 1992, 12 otters

A total of 137 river otters were released in the park. Although river otters may be active day or night during all seasons, they are secretive and seldom seen. Crayfish and fish are their primary foods.

were released in Little River, Hazel Creek, and Cataloochee. Most of the reintroduced otters established home ranges within the park, although one is known to have established a home range in the French Broad River outside the park. These releases (1986, 1988–90, 1990–92) were experimental to determine if otters could reestablish a population in the Smokies. Research results were favorable, so a decision was made to reestablish otter populations parkwide. On January 14, 1994, 50 Louisiana otters were released in the following areas: West Prong of the Little Pigeon River, Middle Prong of the Little Pigeon River in Greenbrier, Big Creek, Cataloochee, Abrams Creek, Little River, Twentymile, and Tabcat. On January 25, 1994, another 50 Louisiana otters were released in the following areas: Oconaluftee, Deep Creek, Eagle Creek, Forney Creek, Pilkey Creek, Chambers Creek, and Noland Creek. The 1994 releases brought the total of reintroduced otters to 137 and successfully concluded the reintroduction effort.

Elk

The feasibility of elk restoration in the park had been under consideration since the late 1980s. A feasibility assessment was completed which concluded that the park had sufficient potential elk habitat to justify an experimental release, and a five-year research proposal was developed to answer questions necessary to determine the future of elk in the park. During 2000, park personnel and wildlife veterinarians completed a disease risk evaluation (DRE), a comprehensive assessment of diseases and parasites of concern with precautionary measures to prevent their ingression into the area. A public outreach program was initiated, and a project "Action Plan" and "Environmental Assessment" for the experimental release was released to the public and all interested parties.

Objectives of the experimental release were: (1) determine dispersal and mortality rates and causes of mortality of reintroduced elk; (2) determine whether mortality or post-release movements vary by age, sex, or reproductive status; (3) assess habitat use; (4) evaluate the effects of variable acorn production on elk demography; (5) evaluate negative impacts of elk reintroduction (e.g., damage to native vegetation or agricultural crops, fence damage, highway mortality); and (6) assess the feasibility, methodology, approach, and probability of success of releasing elk to establish a permanent, viable population in the park.

The first 25 elk (13 males, 12 females) arrived from the U.S. Forest Service's Land Between the Lakes in western Kentucky on February 25, 2001. They were placed in a three-acre acclimation pen in Cataloochee Valley, North Carolina, and held for two months to acclimate them to the area. Two weeks prior to their release, elk were processed and instrumented with radio-collars to monitor their movements. They were released on April 2, 2001, when natural foods were more available. In January 2002 an additional 27 elk (19 females, 8 males) from Elk Island National Park in Alberta, Canada, were released in Cataloochee. Since their release, 62 births have been

Elk are cud-chewers or ruminants and have four-chambered stomachs. They lack upper incisors, which are replaced by a cartilaginous pad. Elk shed their antlers annually. Photo by Steve Bohleber.

recorded. Thirty elk have died. A number of calves have died from predation by black bears and coyotes. A few have been euthanized after exhibiting significant neurological problems. Although necropsy results were inconclusive (2001), symptoms exhibited by the animals suggested meningeal worm (*Parelaphostrongylus tenuis*). No additional elk have been brought into the park. Currently, there are approximately 85 elk (including 2007 births), and all animals are monitored daily. Overall adult survival is 88.6 percent. Four adults died during 2003, including one roadkill. Nine births were confirmed and four others were suspected. Predators (coyotes or dogs) took one of the nine confirmed calves early in the summer. Eight calves were known to have been born during 2004. The deaths of four male and three female elk were documented. The primary source of mortality was parasite related.

During 2005, 10 births were confirmed with possibly two to four more belonging to cows in more remote areas. Predation by black bears resulted in only three of the 10 confirmed births surviving, but these calves were female. The 2005 calving season produced more females than each year before. Approximately 75 percent of all calves born from 2001 to 2004 were males. The best season for newborn elk calves occurred in 2006 when 16 calves were born and 14 survived (88%). In 2007, 19 calves were born and 13 survived (68%).

December 2005 was the end of the initial five-year research phase of the project. The study has been extended for an additional two to three years to help researchers gather data needed to better understand the fate of the Smokies elk herd. Reproduction has been hampered because the herd consists of approximately equal numbers of males and females. In normal elk populations, one male breeds with a harem of females. A third elk release consisting of more females is currently under consideration.

Most elk movements have been confined in or near the park: these areas include Balsam Mountain, Oconaluftee, Qualla Boundary, and White Oak, although one female elk (#42) was located as far as Glenville, North Carolina. She was finally captured in late March 2005 and returned to Cataloochee after a three-year absence. Three male elk moved from Cataloochee to Tennessee. Bull #22 was captured twice in Cocke County and moved back to the park. Another bull moved to Cosby, Tennessee and after a three-year stay moved back to join other elk near Cataloochee. In September 2006, bull #81 moved to the vicinity of Newport, Tennessee.

In the Smokies, bull elk possess fully developed antlers during the fall and winter months. They are used for defense and for competing for the right to breed when the rut (breeding season) begins. New antlers begin growing in the spring as soon as the previous year's antlers have fallen off. Stimulated by increasing levels of testosterone, antlers may grow as much as an inch a day during the spring and summer months. During their development they are covered by a soft layer of velvet that contains blood vessels that carry blood and minerals to the developing bones, which are living tissue. This rapid growth continues for about four months until the antlers reach full size. About August, a bull's antlers will fully mineralize and the new hardened bone will emerge. As the antler hardens, blood flow ceases and the velvet begins to fall off, a process assisted by the bull rubbing its antlers against trees. As a bull ages, its antlers, which consist primarily of calcium and phosphorus and may weigh up to 40 pounds, will continue to grow larger each year if he is in good health and there is an abundance of nutrient-rich food. Larger racks are typically a sign of mature bulls in good health. Bulls will carry these antlers throughout the fall rutting period and until about March. As testosterone levels drop in early spring, the bond between the antler and pedicle weakens, and the antlers fall off. As the days begin to lengthen, testosterone levels begin increasing, and the development cycle begins anew.

Red Wolf

The red wolf is a small, slender, long-legged canid intermediate in size between the gray wolf and the coyote. It is not a pack-structured species, but a highly family-oriented animal, with an adult pair and their offspring making up the basic unit. The red wolf feeds primarily on small, easily caught mammals, but will also take larger animals such as deer when the opportunity presents itself.

In order to prevent the species from becoming extinct, the U.S. Fish and Wildlife Service (USFWS) undertook a captive breeding program in the 1980s. As part of this program, two pairs of red wolves were brought to the park and housed in acclimation pens in January 1991. On April 27, 1991, one female gave birth to a

Although the experimental reintroduction of the red wolf was not successful in the park, the wild population is doing well on the Alligator River National Wildlife Refuge on the coast of North Carolina.

litter of three females and two males. This brought the total known red wolf population to 166.

On November 12, 1991, the two adults and two of their female pups were released in the Cades Cove section of the park. This was an experimental release to determine if the Smokies would provide suitable habitat for the permanent reintroduction of additional red wolves later. The animals gradually expanded their territory to include all of the open area of the cove and the surrounding wooded lands. All were radio-collared and were recorded feeding on grouse, woodchucks, rabbits, raccoons, a juvenile black bear, and deer. The wolves may kill deer as well as feed on deer that are already dead or injured.

In January 1992 the adult male had to be recaptured because after many years in zoos, he was too tame to adjust to life in the wild. In August, his mate was recaptured and both were transferred to the Alligator River National Wildlife Refuge in eastern North Carolina. Their two female offspring were recaptured, paired with males, and rereleased.

On April 25, 1992, the second wild female gave birth to six pups in her breeding pen. Two pups died within 10 days, but the others—three males and a female—were healthy and survived. This family group was released in October 1992 in Cades Cove. An additional pair with four pups was released in the Tremont area in the fall of 1992. A total of 37 red wolves were introduced into the park between 1991 and 1998. Eight litters consisting of 33 pups were born in the park, but only two pups were confirmed to have survived into the fall. Many wolves left the park presumably in search of prey, and some of those that remained succumbed to disease, parasites, and starvation. The lack of survival of wild litters, and the difficulty of keeping wolves within the park caused the USFWS to terminate the Smokies project in October 1998.

Peregrine Falcon

Widespread use of DDT and perhaps other pesticides after World War II gradually eliminated this species as a breeding bird throughout the eastern United States. The last successful nesting recorded in the park was in 1943 and in the state of Tennessee in 1947. The process of destruction began with small organisms being sprayed or eating sprayed vegetation. At each succeeding step in natural food chains, the poison became more concentrated, since each level of organisms consisted of fewer animals than the preceding level—the one on which it feeds. Predators at the end of food chains, such as the peregrine, got the most concentrated doses. In the case of the peregrine falcon and a number of other birds, including the bald eagle and brown pelican, enough poison was ingested to reduce calcium production, causing the birds to lay abnormally thin-shelled eggs which broke or gave the embryo inadequate protection. Unfortunately, no park is an island; a wide-ranging migratory bird like the peregrine cannot be completely protected in parks. Fortunately, the use of DDT has been banned in the United States and some other countries, thus allowing populations of eagles, brown pelicans, and peregrine falcons to once again be able to successfully produce offspring.

From 1984 to 1986, 13 peregrine falcons were acclimated (hacked) and released on Greenbrier Pinnacle in the park. Although peregrines were sighted each year, 1997 was the first year that a pair successfully nested in the park since 1942. They

Peregrine falcons, also known as duck hawks, can be distinguished in flight by their long, pointed wings, their compressed tail, their bold head pattern, and their powerful flight. They feed on many species of small birds, most of which are killed in the air.

produced three chicks (two males, one female) that fledged between July 2 and July 4. Seven peregrine falcons (two adults and five chicks) were observed at the nest site on May 31, 2000. The year 2001 marked the fifth consecutive year that peregrine falcons nested successfully at Peregrine Ridge near Alum Cave Bluff. On May 5, 2001, three young falcons were observed in the original nest, bringing the total over five years to 16 chicks fledged. Although four chicks were observed in the nest in 2002 and two were observed in 2003, it is not known whether they all fledged. Two nest sites were monitored during 2004 and 2005, but nesting success could not be confirmed. Unfortunately, there was no successful breeding during 2006. One of the two nests was not occupied; the young produced in the second nest succumbed before they could leave the nest. Although unconfirmed, one nest was reported to have fledged young in 2007 (Paul Super, personal communication).

Barn Owl

The barn owl is considered an occasional permanent resident of the park. The last observation of barn owls in the park was recorded by Arthur Stupka in 1947. The park library, however, contains two observations from 1975—one in Chimneys picnic area (January 22 at 11:00 p.m.) and the other between the Cosby Ranger Station and Cosby campground (November 21 at 12:30 a.m.)

In 1998 the park initiated a project to augment barn owls in Cades Cove and the surrounding area. Seven young barn owls were banded and placed in an abandoned barn in Cades Cove; however, the fate of these birds is unknown. In 1999, five juvenile barn owls were banded and placed in the abandoned barn; four of these owls were fitted with radio transmitters and located throughout the summer. In 2000, six barn owls were received from the Knoxville Zoo and placed in Cades Cove. There have been no further reintroductions.

The barn owl is relatively slim and long-legged. In flight it appears to have an enormous head, and the bird looks snow-white from below. Photo by Steve Bohleber.

CHAPTER 9

NATURAL HISTORY RESEARCH IN THE PARK

Big animals, little animals, plants—right down to the sea itself. We need them, not just for their own sake, but because all this has to be here for everybody forever. There is no single, simple solution to the problems of continued existence for anything. Any answers are as complicated as life itself. Only one thing is certain: if we are to preserve our environment and save this priceless wildlife, we need much, much more knowledge.

—Harry Butler (1977)

Natural history research in the area now encompassed by the park spans a period of approximately 215 years from Andre Michaux's visit in the 1790s (chapter 3) to the dedication of the Science Center at Twin Creeks in 2007 (chapter 12). During this time, scientists from a wide variety of disciplines have visited the park area and have contributed to our knowledge of its natural history.

Since 1935 three individuals have been primarily responsible for overseeing natural history research in the park: Arthur Stupka, Don De Foe, and Keith Langdon. Among their many other duties, they were primarily responsible for building and overseeing the park's extensive natural history collections. Keith Langdon has been one of the key leaders in organizing and directing the All Taxa Biodiversity Inventory as well as overseeing the construction of the Science Center.

Do You Know:
Who was the first park naturalist?
Who conducted the park's first Christmas Bird Count and the year?
Who organized the first spring Wildflower Pilgrimage and the year?
Who were the cofounders of the Smoky Mountain Field School?
What was the original location of the Uplands Field Research Laboratory?

Arthur Stupka

Arthur Stupka was appointed the first park naturalist in Great Smoky Mountains National Park in 1935. He began his National Park Service career as a ranger-naturalist in Yosemite National Park in 1931 before transferring to Acadia National Park in Maine as a ranger-naturalist and park naturalist in 1932, and from there to the Smokies. What he found upon his arrival was the tiny settlement of Gatlinburg and acres and acres of wilderness.

Stupka, who had received a master's degree in zoology from Ohio State University, was a quiet, reserved gentleman whose legacy will be valued by researchers in this park for generations to come. As a teenager in Ohio, he had begun keeping a nature journal in which he recorded the first arrival of birds in the spring, the blooming dates of wildflowers, and other noteworthy natural events in his life. He was introduced to the field of botany at an early age, and it was quite by accident. He was working at an orchard pruning grapes near a road when a woman stopped her car and asked him if he knew there was a course being offered about pruning and trimming at a nearby college. He investigated and decided to take the course. The instructor was impressed with his interest in the subject and, having no family of his own, offered to finance Stupka's college education. That was an unbeliev-

Art Stupka in 1943 with visitors to the park. Photo courtesy of Maryann Stupka.

able gift, as that was the only way Stupka could have continued his education. He told this story often and was forever grateful to his friend. Without the kindness and generosity of his benefactor, plus the thoughtfulness of the woman telling him about the pruning course, he would not have been able to embark on his career as a naturalist.

Upon his arrival in the Smokies, he began a daily journal of natural history observations in the park which continued for 28 years until his retirement on March 27, 1964. His journal entries have been an invaluable source of information, not only for data that he extrapolated when writing his own books, but also for researchers such as myself.

The first entry in his journal, dated October 14, 1935, reads as follows:

on Oct. 10, 1935, I left Acadia Nat'l Park where, for almost 3 years I had served as Park Naturalist (jr. grade), and journeyed to my new assignment (also promotion) to become Park Naturalist (Assistant grade) in Great Smoky Mts. Nat'l Park (headquarters at Gatlinburg, Tenn.). Went from Bar Harbor, Me., to Washington, D. C. via R.R, and from there I drove a gov't car to Great Smokies—arriving in Gatlinburg, Tenn., on the evening of Oct. 14, 1935. Hereafter, in these journals, the abbreviation *G.S.M.N.P.* will refer to *Great Smoky Mountains National Park*. Locations within the park will be given due to its large area.

His first observations in the park follow:

Oct. 14–20—G.S.M.N.P. (vicinity of Gatlinburg)
What a change from Acadia! the abundant broad-leafed evergreens (rhododendron, laurel, holly, etc.) are especially noticeable here. The mt. slopes are near their height of color—this color does not have the brightness and intensity of the woodland colors in Maine, even tho' the number of trees and shrubs which turn color is much greater (i-e species) than in the north. Altho the *red maple* is a common enough tree here, it is certainly not as abundant as in Maine, and it is certainly not as vivid a red in color. Here the many species of *oaks*, the *sorrel, red maple, gums, sumac, dogwood,* and others turn various shades of russet and red and make a real showing, but not a vivid one.

Tulip trees have lost a wealth of golden leaves by mid-October. The robber-breezes rifle the leaves in great style at this time.

Andropogon grass is abundant in many fields here now arrayed with lovely silken white plumes arranged along its buffy stem. A stand of this grass seen in the right light is a beautiful sight—the seed plumes appearing silvery.

Witch-hazel in full blossom now—later here than in Acadia. Asters and some *goldenrod* in flower, also some *blue violets*.

Saw a number of *fence lizards* (*Sceloporus*) scurrying near the dusty horse trail on Oct. 18.

English Sparrows common here. Perhaps one of the most noticeable birds is the *Carolina Wren*, singing, scolding, or simply asserting itself—very plentiful here.

Heard the fine song of a *winter wren* along a shaded nook beside the Little Pigeon River (near Gatlinburg) on Oct. 20.

White-throated sparrows here in numbers—occasionally uttering their half-hearted autumn song. Watched 2 of them feeding on the red fruits of dogwood. Flicker, downy here.

Goldfinches, bluebirds, killdeer, cardinals, kingfishers, phoebes, bobwhite, mockingbird, waxwings, starlings, here.

Monarch butterflies are common—frequently to be seen fluttering by.

Clematis vine adorned with silvery-gray clumps of feathery plumes—a common roadside plant here.

Saw a young *oppossum* [sic] run across the road between Gatlinburg and Sevierville on the night of Oct. 15.

Chipmunks appear to be common here. Saw one gray-squirrel.

Found a dead *blacksnake* and a *watersnake* in the road near Gatlinburg.

A goodly *insect chorus* in the night.

The first naturalist quarters were located in the old Pi Beta Phi House in the Sugarlands in April 1936. In June 1936 the Bruce Keener house near the mouth of Fighting Creek was acquired by the park, and beginning in December was used as park naturalist offices and temporary museum until the present Park Headquarters Building was completed in 1940. In 1941, the second floor of the Administration Building was completed and was used to house a collection of mountain culture artifacts, herbarium, other collections, and office space for ranger-naturalists. Sixty collecting permits were issued in 1941. Some early research supervised by Stupka included:

1935	Dr. Carl Hubbs made an extensive survey of fishes.
1940	A. C. Cole of the University of Tennessee published a guide to ants of the park.
1941	Dr. H. R. Hesler of the University of Tennessee prepared a list of fungi which included 1200 species.
	Herman Alva of the University of Tennessee made a study of algae.
	Gunnar Digelius collected 206 species of lichens in the park.
1942	"A Preliminary List of Trees and Shrubs" was prepared.

	Seventy-four species of fish were recorded. Seventy-four species of reptiles and amphibians were recorded.
1946	The first fossils were discovered in the park by USGS geologist R. B. Neuman. The fossils were brachiopods, cephalopods, and trilobites.
1951	R. R. Dreisback of Midland, Michigan, presented 100 insect specimens to the park. Fred Calle began a study of azaleas on Gregory Bald. R. B. Neuman published an account of "the Great Smokies Fault."
1952	R. H. Whittaker published his study of summer foliage insect communities and donated 150 insect specimens to the park. A. C. Cole of the University of Tennessee published a checklist of ants of the park.

Stupka is credited with a pioneering effort to catalog the park's plants and animals. As one writer noted: "In a time when many, many people are expending untold hours of effort in the attempt to update knowledge regarding biodiversity in the Smokies, we are humbled by the contemplation of the enormity of Stupka's accomplishments." His thousands of handwritten observations have given scientists an invaluable baseline for measuring environmental changes in the Smokies. He assisted in mapping many of the trails, searched out the tallest and biggest trees, the abundant waterfalls, and the rarest wildflowers. Today his work forms the foundation of the All Taxa Biodiversity Inventory (ATBI) (see chapter 10). In his book *Strangers in High Places*, Michael Frome wrote, "He [Stupka] was on more intimate and knowing terms with nature in all corners of the park than any other man during his time."

After having been in the park only a short time, he conducted the park's first Christmas Bird Count on December 19, 1935, finding 39 species in the Cades Cove area. The Knoxville Bird Club joined him in 1937, and two dozen birders helped with the count. Along with Dr. Jack Sharp of the University of Tennessee, Stupka began the annual Spring Wildflower Pilgrimage in 1950. The 58th Pilgrimage will be held in April 2008.

Stupka spent four years hiking, building a natural history collection, and making connections with scientists before he offered a single public hike or evening program. In June 1937 Dr. Stanley Cain and park naturalist Stupka organized the first Nature Trail in the Smokies. It was called the Greenbrier-Brushy Mountain Nature Trail. On July 5, 1939, Stupka began a regular program of naturalist-guided trips that consisted of two-hour, half-day, and all-day trips to such places

Table 9.1. New species and new genus of invertebrates named in honor of Arthur Stupka.

Species	Common Name	Year	Collector/Author
Castianeria stupkai	Spider	1940	Dr. W. M. Barrows
Limnophila stupkai	Cranefly	1940	Dr. C. P. Alexander
Erythroneura stupkaorum	Leafhopper	1945	Dorothy J. Knull
Phyllobaenus stupkai	Beetle	1949	Dr. J. N. Knull
Laelaps stupkai	Mite	1971	Dr. Donald W. Linzey, Dr. Douglas A. Crossley
Stupkaiella (new genus)	Fly	1973	F. Vaillant
Nesticus stupkai	Spider	1989	J. M. Harp, R. Wallace

as Laurel Falls, Rainbow Falls, Andrews Bald, and Charlies Bunion, with overnight hikes to Mount LeConte. He scheduled talks in hotel lobbies on the evenings prior to his walks and showed slides previewing what hikers would see the following day. He later commented, "I was lucky to be first on the scene. I was my own boss."

A friend once commented: "He was a sharing sort of guy. The gift of a great park naturalist. He loved it and they loved him."

His daughters, Maryann and Carolyn, developed a greater appreciation for nature by accompanying their father on some of his walks and attending some of his talks. Carolyn remembers being carried piggyback by her father to Mount LeConte. To Maryann, a bug in the house was one to be carried outside to its place in nature.

His grandson, Teddy Murrell, said it best: "He loved to take somebody out on a nature walk who didn't know a trillium from a chickadee, and share his knowledge and enthusiasm about everything they found."

Over the intervening years, scientists from a variety of disciplines showed their respect for Art Stupka by naming new species after him. A total of six species and one genus possess his name (Table 9.1).

In 1960 the decision by the National Park Service to emphasize "entertaining" visitors resulted in a deemphasis on natural history. In October 1960, in large part due to this new focus, and after completing 25 years in the capacity of park naturalist, Stupka filled the newly created position of biologist in Great Smoky Mountains National Park. In this capacity he could avoid becoming involved with public relations and money. He remained in this position for the last four years of his career, saying, "I retired because I didn't want to have anything to do with financial things."

Stupka once wrote:

> Many people envy me my job and well they might. For it is my duty to make the visitor better acquainted with a truly grand area—the Great Smoky Mountains National Park. Hiking trips are conducted to exceptionally attractive regions of the

Art Stupka in 1960. Photo courtesy of Maryann Stupka.

park, and in the course of these trips informal explanations of trailside objects are given. Illustrated talks on phases of the natural history of the Smokies are given and information pertaining to the plant and animal life is compiled. In preparation now are various lists of floral and faunal groups, and eventually these will be available to all who are interested. Plans for the realization of natural history and pioneer culture museums are also well under way. Biologists who visit here and who have made a specialty of certain groups which are represented in this Park are given every possible cooperation, for their findings are of mutual benefit—and it's going to take a great many such students a mighty long time before we can say that we are pretty well acquainted with all that is here.

I am doubly fortunate, for my out-doors laboratory is not only a naturalist's paradise, but the sort of people with whom I come in contact are the finest in the world.

In addition to writing a number of leaflets and other nature publications, Stupka is the author of *Notes on the Birds of the Great Smoky Mountains* (University of Tennessee Press, 1963), *Trees, Shrubs, and Woody Vines of Great Smoky Mountains National Park* (University of Tennessee Press, 1964), and *Wildflowers in Color* (Harper & Row, 1965). He is the co-author (with James E. Huheey) of *Amphibians and Reptiles of Great Smoky Mountains National Park* (University of Tennessee Press, 1967). In 1965, he asked my wife and myself, both mammalogists, if we would compile all known mammal data for the park. We agreed and published

Mammals of Great Smoky Mountains National Park (University of Tennessee Press, 1971). The dedication reads: "To Arthur Stupka—in recognition of his immeasurable contribution to the knowledge of the natural history of the Great Smoky Mountains National Park."

During many of our visits to Art and Margaret's home on Buckhorn Road, we would sit outside on the patio in the dark waiting to hear a thump on the tree containing a bird feeder. Art would immediately train his flashlight on the feeder, and we could see one or more southern flying squirrels feeding on the sunflower seeds. Margaret was a photographer and had her master's degree in Botany, both of which proved to be assets to her husband's research.

Upon his retirement in 1964, Art Stupka was presented with the Meritorious Service Award, the second highest honor an employee of the Department of the Interior can receive. Secretary of the Interior Stewart L. Udall stated:

> His work has provided visiting scientists from throughout the world with inspiration and information. Millions of visitors are benefitting from his contributions to the planning and development of interpretive exhibits in visitor centers of the park. The heritage he leaves in the twenty-eight unpublished volumes, "Nature Journals of Great Smoky Mountains National Park," is of inestimable value for scientific reference. His cooperation and collaboration with scientists has increased biological knowledge and nature appreciation. Publications he authored interpret the natural wonders of the park and have led to better popular and scientific appreciation of park resources. He has made a major contribution to good administrative management of park protection, interpretation, and development. He has literally made thousands of people "see" nature for the first time. His efforts have done much to advance the cause of conservation. In recognition of his outstanding service as a park naturalist and biologist, his contributions to scientific research and knowledge, and his ability as an interpreter of nature in Great Smoky Mountains National Park, the Department of the Interior grants to Mr. Stupka its Meritorious Service Award.

In 1975 Art Stupka was named an Honorary Member of the Association of Interpretive Naturalists Inc. In October 1987 he was elected a Fellow of the Tennessee Academy of Science, an honor accorded to those members of the academy who have demonstrated an outstanding record of scientific accomplishment in their particular field and who have actively participated in the affairs of the academy. In February 1998, at the age of 92, he was presented with the Tennessee Ornithological Society's highest award. He was only the second person in the society's 82-year history to receive the award.

The Hemlock Inn in Bryson City, North Carolina, was a special place for Stupka. He would go there in the spring, fall, and sometimes several weeks each season. From 1983 until the mid-1990s, Stupka spent the winter months in Florida and the summers in the Smokies. Every spring when he returned to the mountains, he would spend a week at the Inn giving illustrated talks and leading nature walks. He had a yearly following of people coming to the Inn who shared his interest in nature. A trail is named in his honor. During his last year at the Hemlock Inn when he was 90 years of age (1995), my wife and I visited with him. He rode in the passenger seat of our car as we led a car caravan along the Road to Nowhere outside Bryson City. He would direct me to stop at particular sites so that everyone could get out of their cars and be informed about what they were seeing. This was a trip that I will never forget.

Arthur Stupka passed away on April 12, 1999, at the age of 93.

Don De Foe

On November 1, 1963, Don De Foe began his duties as park naturalist in the Smokies. His NPS career had begun as a park ranger, and later park naturalist, at Lake Mead National Park in Nevada in October 1959. Except for a six-year position as a supervisory park naturalist on the Blue Ridge Parkway, his entire career was spent in the Smokies, where he served for 24 years as assistant chief of interpretation before becoming museum curator in September 1996. An authority on the natural history of the park and highly skilled in field biology, Don De Foe was one of the last "old-time naturalists" in the best tradition of the National Park Service. In an age of increasing specialization, he strived to know and, if possible, understand all the parts of one of the most complex natural systems in the temperate world. He also marveled at and loved its beauty. Don was well known to hundreds of scientists over the years, and collaborated on manuscripts with them. He was also a valued

Don De Foe. Photo courtesy of Jay De Foe.

and helpful source of information for his many park colleagues, friends in the community, and visitors alike. He was extremely well organized and meticulous about his work; his counsel was very well respected and sought on many issues.

As museum curator, he was responsible for preserving and documenting the park's internationally known biodiversity. He organized and upgraded the natural science collections, which already included many hundreds of specimens that he had collected. Some of his collections represented species that were rare, endemic, or undescribed to science. When popular focus was on the larger, "charismatic" species of animals and plants, Don helped us to see the ecological wisdom of protecting even the most seemingly obscure species. He assumed leadership of the Christmas Bird Count from 1972 until 1995. He was a member of the team selected to go to Costa Rica in 1998 to plan the Smokies' All Taxa Biodiversity Inventory. My first contact with Don came in June 1964, when I became the first seasonal park ranger-naturalist that he hired. Our friendship continued for the next 40 years.

Don, who like Art Stupka was naturally shy, quiet, and unassuming, was instrumental in overseeing 20-plus seasonal naturalists each year during the 1960s and 1970s. Although he did not write any books, he used research findings as a theme to reconfigure the natural history exhibits at Sugarlands Visitor Center. While he was extremely knowledgeable about all aspects of the park's natural history, his main focus was on birds and insects.

Don was a cofounder (with the University of Tennessee) of the Smoky Mountain Field School, which brings the public together in the field with scientists to do one-day intensive learning about a single aspect of natural resources. He also worked to expand the Wildflower Pilgrimage in April each year. He received numerous special achievement awards and the Department of Interior's Award for Superior Service in 1997.

Don De Foe passed away on February 2, 2003, at the age of 68.

As noted earlier in this chapter, Arthur Stupka filled the newly created position of park biologist from October 1960 until his retirement in 1964. Scientific research in the park slackened after 1964, and the position of park biologist has remained vacant. In September 1974 Susan Bratton, a recent Ph.D. graduate of Cornell University, was recruited to fill the position of coordinator of research. Bratton had been working on her dissertation on the effects of the European wild boar in the park since 1972. Her appointment was not without controversy. Although park superintendent Vincent Ellis did not think the park needed a biologist, regional chief scientist Ray Hermann charged her with creating the Uplands Field Research Laboratory to "conduct high-priority research in the Great Smoky Mountains and other upland national parks in the Southeast region of the National Park Service and to monitor environmental change and conduct other research projects which address threats to resources or pose management concerns."

In her book *The Wild East*, Margaret Lynn Brown detailed the obstacles and disdain faced by Bratton and her Uplands crew of researchers, many of whom were graduate students. For example, superintendent Ellis housed the research facilities in a collection of run-down trailers at Tremont, 16 miles from park headquarters. Bratton was assigned a beat-up pickup truck that frequently broke down. She had no desk, no lamp, and had to use a borrowed typewriter. But research was accomplished and management reports, master's theses, dissertations, and scholarly articles were produced both by Uplands scientists and outside researchers—Susan Bratton, Peter White, Michael Pelton, Christopher Eager, Dan Pittillo, Charlotte Pyle, John Peine, Nicole Culbertson, Mary Lindsay, Mark Harmon, James Renfro, myself, and many others.

One major concern of Bratton and her researchers was the effect of overuse on the environment from such activities as hiking, horseback riding, illegal camping, and the use of motorized vehicles for maintenance and ranger patrols in the backcountry. Superintendent Ellis was not sympathetic to Bratton's concern of "lack of management" in the park.

In 1975 Boyd Evison was appointed superintendent and moved into the superintendent's home at Twin Creeks (see chapter 12). Within the first few weeks of his tenure, he recognized the value of the research laboratory, but was appalled by the condition of its facilities, its distant location, and the fact that Bratton had been effectively excluded from the park's management team and its decision making. To reconcile these conditions, Evison moved his family into the district ranger's home and allowed the Uplands Field Research Laboratory to move to Twin Creeks. Evison not only supported Bratton's work on trail erosion and stream siltation from overuse, but supported publication of a brochure explaining Uplands' function in research coordination, collecting baseline data, monitoring research, project research, and management application. Evison and Uplands reversed a long history in the park of management without science.

From 1977 to 1980 Gary Larson served as coordinator of research, with Bratton's title being research biologist (botanical). Under the direction of Bratton and Larson, the Uplands laboratory conducted historical research on the conditions of balds and former agricultural uses, recommended policy that protected native species from such introduced species as the European wild boar and the rainbow trout, and began a management report series (Research/Resources Management Series) to put research into print quickly and make it more accessible to nonscientists.

In 1980 Bratton accepted a position at the University of Georgia to direct the NPS Cooperative Unit. Peter White served as acting coordinator of research from 1980 to 1981 until John Peine, a sociologist, was hired in 1982 to fill the position of coordinator of research. Peine continued to head the Uplands Laboratory through 1993.

Joe Abrell filled the position of chief of resource management and science on June 29, 1988. His responsibilities included coordinating the natural and cultural resources programs as well as the Uplands Field Research Laboratory. He was succeeded by Larry Hartmann (June 1999–June 2005) and the current chief, Nancy Finley, who began her duties in February 2006.

Under the administration of President Ronald Reagan and his Secretary of the Interior James Watt, Uplands Field Research Laboratory disappeared into the Science Division, which reported more directly to the superintendent. During this time, the National Park Service shifted field research to contracts by university-employed professors, effectively eliminating in-park research.

The next secretary of the interior, Bruce Babbitt, formed a new agency, the National Biological Service (NBS), which was later renamed the National Biological Survey, into which he transferred almost 2000 scientists from in-house research stations and cooperative units of the National Park Service and the U.S. Fish and Wildlife Service. This new agency was charged with providing biological inventories, monitoring, research, development, and technical assistance. Later, the NBS became the Biological Resources Division (now "Discipline") under the U.S. Geological Survey.

Keith Langdon came to the park as vegetation management specialist in 1985. In 1993, he became inventory and monitoring coordinator for the new Inventory Monitoring Program. This program operated under the auspices of the U.S. Geological Survey for two years before being transferred to the National Park Service. The Inventory and Monitoring Branch was located in the facilities formerly occupied by the Uplands Field Research Laboratory at Twin Creeks, which was renamed Twin Creeks Natural Resources Center. It has provided offices and facilities for inventory and monitoring, air quality control, forestry and exotic species, and the All Taxa Biodiversity Inventory (see chapter 10). Personnel are not engaged in basic research, but in inventorying and monitoring activities. Current personnel include an entomologist (Becky Nichols), a botanist (Janet Rock), an ecologist (Mike Jenkins), the museum curator (Adriean Mayor), a data manager (Michael Kunze), a GIS specialist (Ben Zank), and Chuck Parker (research aquatic biologist), who represents a one-person field station for the USGS. Parker, who began his work in the Smokies in February 1986, had served as interim director of the Uplands Laboratory during its final year of existence.

The concept of an All Taxa Biodiversity Inventory was conceived by Keith Langdon, Chuck Parker, and John Pickering (University of Georgia) while sitting on the front porch of the Twin Creeks Natural Resources Center during the fall of 1997. The discussion centered around the fact that the resources that Langdon and Parker were charged with monitoring were largely unknown. As Langdon once said, "If you inherited a hardware store from your father, the first thing you would do is to take an inventory of what was in it." With the assistance of Pickering and David Janzen, a University of Pennsylvania professor who had begun the first ATBI in Costa Rica, the park's first ATBI conference was organized and held in December 1997. The rest is history (see chapter 10).

It is interesting to note that several persons associated with the Uplands Laboratory, either as Uplands scientists or outside researchers, have continued their careers and research activities with the National Park Service and are currently playing major roles in the Resource Management and Science Division and/or the All Taxa Biodiversity Inventory. For example, Steve Moore is currently supervisory fishery biologist; Jim Renfro is currently supervisory air quality specialist; Chuck Parker is currently research aquatic biologist; and Peter White, currently director of the North Carolina Botanical Gardens, is chairman of the DLIA Board of Directors. Since 1999 I have served as chairman of the Mammal Taxonomic Working Group (TWIG) for the ATBI.

Many other researchers and their students have worked on natural history-related projects in the park over the years. While it is not possible to recognize every individual in a book such as this, their combined efforts have resulted in our current knowledge base of the park's natural history.

The Twin Creeks Science Center, completed in 2007, will be the next chapter in the park's science endeavors (see chapter 12). These facilities will serve as a base for future science activities for many years to come.

CHAPTER 10

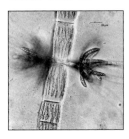

ALL TAXA BIODIVERSITY INVENTORY

The universe would be incomplete without man; but it would also be incomplete without the smallest transmicroscopic creature that dwells beyond our conceitful eyes and knowledge.

—John Muir (1901)

Purpose and Goals

The All Taxa Biodiversity Inventory (ATBI) is a concentrated effort to identify every species of plant and animal living within Great Smoky Mountains National Park within a relatively short time frame. It is the first comprehensive biological inventory ever undertaken in North America. The ATBI in the Smokies was conceived in late 1997, in part as a prototype for other reserves. Basic approaches for sampling were worked out by late 1998, and funding was sought for a pilot program which began in fall 2000. Funding for the pilot years came from NSF, USNPS, USGS, the Friends of the Smokies, and the Great Smoky Mountains Association as well as from individual donors and some community corporate support. The goal of the ATBI is not just to compile lists of what occurs in the park, but to (1) discover the parkwide distribution of each species, (2) determine the relative abundance of each species, and (3) gather data on the seasonality and ecological relationships of all species in the park. Information resulting from this project will help park managers make decisions necessary to protect critical natural resources from a growing list of threats. Over 200 scientists are contributing their expertise by working in the park. Many other scientists contribute their time in identification of species, but may never visit the park.

The ATBI has generated a strong interest in biodiversity, natural history,

Do You Know:
The purpose and goals of the ATBI?
How many new species have been found in the park since 1998?

taxonomy, and conservation. The involvement of nonscientist citizens and students as "citizen scientists" has been significant. For example, our neighbor who is a high school junior undertook a study of snail behavior at different elevations in the park. She documented through notes and photographs the movements and behavior of snails during 24-hour periods.

Discover Life in America Inc. (DLIA) is a nonprofit organization founded in 1998 that works under a cooperative agreement with the National Park Service and Great Smoky Mountains National Park to conduct the ATBI. Receiving support from the Great Smoky Mountains Association, Friends of the Smokies, as well as individual and corporate donors, DLIA administers grants to scientists; organizes volunteers; coordinates development, marketing, and public relations; and assists park partners with education programs. The Web address for Discover Life in America is www.discoverlifeinamerica.org or www.dlia.org.

Through 2005, six National Science Foundation grants (specific to particular subprojects) have been competitively awarded to collaborating scientists. "Mini-grants" totaling over $300,000 have been competitively awarded by DLIA and each year have attracted three to 10 times as much funding and in-kind services as the mini-grants themselves. Efforts are currently under way to organize an alliance of ATBI parks and reserves to promote biodiversity research, funding, education, communication, and exchange of data.

ATBI Results

Estimates of the number of nonmicrobial species in the park vary from about 50,000 to 100,000 or more. About 15,450 are currently known. Since the beginning of the ATBI in 1998 through December 2007 the following numbers of new species have been found:

> 861 species new to science
> 5,203 species new to the Great Smoky Mountains National Park.

Species new to science are ones that have never been formally identified and described. New records for the park are species that were previously known to science but until now had not been found in the Smokies. Numbers in Table 10.1 change rapidly as more research results come in. Many of the species in the "New to Science" category have not yet been published, but have been verified by professional taxonomists.

What is very apparent from the above list is that only 11 of the new park records are of terrestrial vertebrates: two amphibians (eastern spadefoot toad and mole salamander), two turtles (common musk turtle and Cumberland slider), six birds (short-eared owl, redhead duck, chestnut-collared longspur, harlequin duck, short-billed dowitcher, and mourning warbler), and one mammal (evening bat).

Table 10.1. New species recorded for Great Smoky Mountains National Park, 2000–2007.

TAXON	Old Records (prior to ATBI)	New to Park (since ATBI began, 1998)	New to Science	Total New
Microbes				
Archaea	0	0	44	44
Bacteria	0	179	270	449
Protozoa	0	33	2	35
Cnidaria				
(jellyfish, hydra)	0	2	0	2
Bryozoa				
(moss animals)	0	1	0	1
Slime Molds	100	150	21	271
Algae	354	528	70	952
Fungi	2,102	583	37	2,722
Lichens	394	154	15	563
Plants				
Vascular	1,598	62	0	1,660
Nonvascular	482	9	0	491
Nematomorpha				
(horsehair worms)	2	3	0	5
Mollusks				
(snails, mussels, etc.)	108	64	7	179
Annelids				
Aquatic oligochaetes	1	27	2	30
Earthworms	13	8	4	25
Platyhelminthes				
(flatworms)	11	8	1	20
Nematodes				
(roundworms)	6	6	1	13
Tardigrades				
(waterbears)	3	56	14	73
Arachnids				
spiders	224	240	41	505
mites	22	55	31	108
ticks	7	4	0	11
harvestmen	0	12	0	12
Crustaceans				
(crayfish, copepods, etc.)	14	66	27	107
Chilopoda				
(centipedes)	20	9	0	29
Diplopoda				
(millipedes)	3	28	1	32
Pauropoda				
(pauropods)	5	27	13	45
Symphyla				
(symphylans)	0	0	2	2
Protura				
(proturans)	11	5	10	26

Table 10.1. (cont.)

TAXON	Old Records (prior to ATBI)	New to Park (since ATBI began, 1998)	New to Science	Total New
Collembola				
(springtails)	130	51	47	228
Diplura				
(diplurans)	6	2	5	13
Microcoryphia				
(jumping bristletails)	0	2	1	3
Ephemeroptera				
(mayflies)	125	6	4	135
Odonata				
(dragonflies, damselflies)	61	32	0	93
Orthoptera				
(grasshoppers, crickets, etc.)	70	35	0	105
Orthopetroids				
(cockroaches, mantids, walking sticks)	6	6	0	12
Plecoptera				
(stoneflies)	108	1	5	114
Psocoptera				
(barklice)	18	49	8	75
Pthiraptera				
(lice)	8	37	0	45
Hemiptera				
(true bugs, hoppers, etc.)	236	213	4	453
Thysanoptera				
(thrips)	0	46	0	46
Neuroptera				
(lacewings, antlions, etc.)	12	31	0	43
Coleoptera				
(beetles)	887	1,223	34	2,144
Mecoptera				
(scorpionflies)	17	2	1	20
Siphonaptera				
(fleas)	17	9	1	27
Diptera				
(flies)	495	170	41	706
Trichoptera				
(caddisflies)	156	79	4	239
Lepidoptera				
(butterflies, moths, skippers)	891	402	74	1,367
Hymenoptera				
(bees, ants, etc.)	209	471	19	699

Table 10.1. (cont.)

TAXON	Old Records (prior to ATBI)	New to Park (since ATBI began, 1998)	New to Science	Total New
Vertebrates				
fishes	71	6	0	77
amphibians	41	2	0	43
reptiles	38	2	0	40
birds	237	6	0	243
mammals	65	1	0	66
Total	9,384	5,203	861	15,448

Taxa table prepared by Becky Nichols, NPS.

This discrepancy was fully anticipated since previous work in the park has concentrated on the larger and most highly visible forms. Although most of the new species are small in size, they are often big in importance to the health of ecosystems. They play crucial roles in cycling of nutrients, as predators or prey, or even improving the vigor of trees and other plants by fixing nitrogen in the soil.

One good example are diatoms, a group of microscopic algae. As the base of the food chain, diatoms are cosmopolitan in their abundance within almost all types of aquatic habitats. Their global biological importance is dramatic. Both freshwater and marine diatoms are estimated to remove nearly half of all the carbon dioxide (a greenhouse gas) from the earth's atmosphere by photosynthesis. That's more than all the tropical rain forests, temperate forests, and grasslands combined! In addition, diatoms produce globally at least 25 percent of the oxygen we need to breathe. Diatoms are used as environmental indicators in streams and rivers to assess a watershed's biological integrity. Many state and federal Environmental Protection

The green alga *Draparnaldia appalachiana* is a park endemic known only from the Cades Cove sink hole. Photo by Rex L. Lowe.

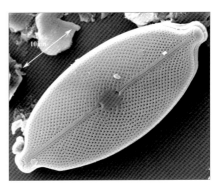

Decussata placenta, a subaerial diatom from the park. Photo by Jennifer Ress and Rex L. Lowe.

Agency staff use diatoms, along with fish and aquatic insects, to biologically monitor streams. Like fish and aquatic insects, individual species of diatoms can signal pollution. Certain species are pollution tolerant, while others are more pollution sensitive. Temperature, light levels, nutrient resources, pH, and toxic materials can also dictate diatom community distribution, and they respond quickly to environmental changes. Diatoms are an excellent biomonitor because of their short generation time, doubling their population about once a day. Due to this short generation time, diatoms are one of the first to recolonize an area after an environmental disturbance.

One of the most interesting and little-known groups of invertebrates in the park are the tardigrades or water bears (Phylum Tardigrada), a sister group to the arthropods. The cigar-shaped body of a water bear consists of four segments, a head, four pairs of stubby legs without joints, and feet with claws. They have a ventral nervous system and a multilobed brain but lack respiratory and circulatory systems. Most adults are less than 1.0 millimeter in length and can just barely be seen with the naked eye. The body of all adult tardigrades of the same species possess the same number of cells, a condition referred to as *eutelic*. Tardigrades occur from the polar regions to the equator and from 6,000 m above sea level in the Himalayas to 4,000 m beneath the surface of the ocean. They feed on plants or small animals.

In the park, tardigrades are found in leaf litter and in the interstitial spaces of sand in the bottom of stream beds. They are especially common in the thin film of water that covers mosses, liverworts, and lichens. This unpredictable type of habitat is periodically wet and dry and may even become frozen. Terrestrial tardigrades can survive drying and environmental extremes by undergoing cryptobiosis, a

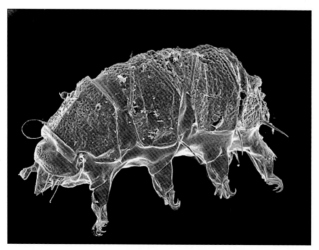

Echiniscus virginicus, one of 74 species of tadigrades (water bears) currently known from the park. Photo by Paul J. Bartels and Nigel Marley.

kind of suspended animation, by converting themselves into a "tun." This involves pulling in their legs to give their body a cylindrical shape and then shutting down all their metabolism. Any other animal exhibiting zero metabolism is considered dead. Only tardigrades regularly return to life again from this state. They perform this miracle by replacing the water in their cells with a sugar called trehalose and glycerol. While in this state, which may last for 10 years or more, they may be more resistant to severe environmental extremes than any other known organism. Thus, cryptobiosis may enhance not only their survival but also their dispersal.

Paul Bartels of Warren Wilson College and Diane Nelson of East Tennessee State University have been studying tardigrades in the park since 2001 as part of the All Taxa Biodiversity Inventory. Prior to this study, only three species were known from the park. The current species list now stands at 75 species, including 70 new park records and 14 probably new to science. One species of particular interest is *Bryodelphax* sp. It is only the second species from this genus to be reported in North America and is likely new to science. The first species was found in western Canada and Wyoming. Thus far in the park, this species has only been found on lichens outside caves in the Cades Cove area, indicating that it may be limestone dependent.

Tardigrades are more or less evenly distributed in soil, lichens, moss, and streams. On trees, the number of species was greater in mosses at breast height than in mosses at the bases of trees. Soil characteristics were found to be major determinants of tardigrade diversity, rather than elevation or slope. Preliminary work on tardigrade response to acidification shows that activity is reduced at pH levels of 3 and 4 with death occurring within five minutes at pH 2.8. Such studies are significant because pHs of 5 and 4 have been reported in park streams. Further stream acidification levels could prove detrimental not only to tardigrades but to entire aquatic communities. Based on species richness estimates, Bartels and Nelson estimate that as many as 96 species may be found to occur within the park. Ninety-six species represents 10 percent of the world's known tardigrades, whereas the park is home to only 6 percent of the world's salamanders. The database of 8,500 entries is the second largest tardigrade database ever created and the only computer/georeferenced database.

An extremely rare lichen was discovered during EIS work on the Ravensford tract by Dr. Tor Tonsberg, a lichenologist from Norway. Repeated searches failed to find this species anywhere else in the park in similar habitat, and the thalli on one tree disappeared during 2004. The lichen, newly described and named *Leioderma cherokeense*, was transplanted to park land in similar habitat. As of September 2005, it was doing well. Until the discovery at Ravensford, only one other *Leioderma* species was known from North America—another rare species from the Puget Sound area in Washington State.

As a result of continuing ATBI investigations, many new species are being added to the park's flora. For example, three new plants were discovered in the park in 2002. Eastern leatherwood is a native shrub that has incredibly flexible twigs that were used by some Native Americans for rope and other cordage. It has a widespread distribution in eastern North America, but is rare south of Virginia. Its discovery represents a new family of plants in the park. Populations of mercury spurge and serrinated skullcap were also confirmed. Twenty-nine vascular plants were added to the park's flora in 2003 and 2004, five were added in 2005, six were added in 2006. These plants are included in Appendices II, III, and IV with a notation as to their year of discovery in the park.

Despite the history of botanical work in the park over the last century, there are still large or easily recognizable species of vascular plants that remain unrecorded here. Wherever comprehensive inventories are undertaken in any region with complex landscapes, it is usually the rare species that are last to be found because of their limited distributions. The ATBI will continue to provide park managers with information on new exotic species that invade as well as on rare native species.

Researchers use a variety of techniques appropriate to their particular group of organisms. In addition to hand collecting, dip nets, seines, and electroshocking are used by fisheries biologists; large pieces of tin are placed on the ground to serve as cover for snakes and lizards; birds and bats are captured with mist nets; mammals are sampled with mist nets, live traps, and pitfall traps; and insects are lured by Malaise traps. A technique utilized by some groups of researchers is a "bio-quest" or "bio-blitz." A bio-blitz is a coordinated sampling effort by groups of experts and volunteers who fan out over the park and collect as many of the appropriate organisms as possible in a specific time period. Discover Life in America has sponsored such sampling efforts for several groups: moths and butterflies, beetles, snails, bats, slime-molds, millipedes, protists, fungi, and bryophytes. At the second Lepidoptera Bio-blitz in June 2002, nearly 30 "leppers" (that's what they call themselves) identified 860 species of Smokies moths, butterflies, and skippers in a 24-hour period. This broke the previous North American record for this type of time-constrained sampling that was set here in the park in 2000 in which 720 species were documented. The 2002 survey located 138 species that were previously unknown in the park, and 51 species that were believed to be new to science.

The total number of lepidopteran species for Great Smoky Mountains National Park is now over 1,300. Besides the record number of species recorded in the 2002 bio-blitz, a significant rediscovery was made. The tawny crescent butterfly (*Phyciodes batesii maconensis*) has declined for unknown reasons over most of its range and is considered a "federal species of concern." It has a subspecies that is only known from the southern Appalachians. This subspecies had not been seen in recent decades, and some experts speculated that it had become extinct. The only park records were from the 1930s, and repeated searches over the last decade had failed to find it. Some populations were located outside of the park very recently, and during the 2002 bio-blitz, a population was finally rediscovered in the park, where it will be studied and protected.

One interesting technique was a tree canopy biodiversity research project during the summer of 2000. Researchers from seven universities including bryologists (liverwort and moss experts), a lichenologist, an ecologist, a mycologist (fungi), a myxomycetologist (slime molds), and a flowering plant systematist served as mentors and as the "ground crew" for a group of student climbers who accessed the tree canopy using the double rope climbing technique to explore and collect myxomycetes, macrofungi, mosses, liverworts, and lichens. They went "where no one has gone before." Their objectives were to: (1) initiate the first survey and inventory of tree canopy biodiversity for myxomycetes, macrofungi, mosses, liverworts, and lichens in GSMNP; (2) compare the assemblages of tree canopy groups of cryptogams with those on ground sites; (3) compare species diversity of the targeted organisms between tree species; and (4) search for undescribed taxa new to science in all of the targeted groups of organisms. Targeted samples were collected from over 160 trees representing over 25 different tree species. These samples were scanned for specimens directly on the bark surface using a dissecting microscope. Bark samples were then placed in moist chambers for culture of the organisms. Identification of lichens from the approximately 3,000 specimens obtained in this initial survey has not yet been completed, but preliminary data indicates that a probable total of between 200 to 300 species were found. Of these, perhaps up to 100 species may be new to the park's current lichen list.

CHAPTER 11

ENVIRONMENTAL CONCERNS

Humankind has not woven the web of life. We are but one thread within it. Whatever we do to the web, we do to ourselves. All things are bound together. All things connect.

—Chief Seattle (1855)

Environmental science is the study of how humans and other species interact with one another and with the non-living environment. Its goal is to develop ways of living that permit all species to survive and prosper into the indefinite future. It uses and integrates knowledge from physics, chemistry, biology (especially ecology), geology, geography, resource technology and engineering, resource conservation and management, economics, politics, and ethics. In other words, it is a study of how everything works and interacts—a study of connections in the common home of all living things.

The greatest threat to many parks today is posed by human activities. In Great Smoky Mountains National Park, wildlife and recreational values are threatened by coal-burning power plants, urban development, and alien species that have moved or been introduced into the park. Polluted air drifts hundreds of miles to kill trees and other plants, acidify soils and waterways, and blur the awesome vistas.

The Smokies are within a day's drive of a third of the U.S. population and play host to over 9 million visitors annually—more than any other national park. Some 2 million people visit Cades Cove every year—more than the total annual visitation to most national parks. People are loving the park to death!

The infrastructure needed to handle the large number of visitors—

Do You Know:
What is considered Public Enemy #1 in the park?
Why there are no longer American chestnut forests?
Where the balsam woolly adelgid came from and when?

roads, picnic areas, campgrounds, water pipelines, sewage facilities, electricity, a campground store—all impact the environment even though their presence and effects are minimized by the National Park Service. The increasing number of motor vehicles contributes to the degradation of the air quality in the park and surrounding areas.

Air Quality

Air pollution exists in three forms: particulate matter, which creates haze and can result in respiratory problems; acid deposition, which impacts both terrestrial and aquatic environments and organisms; and ozone, which impacts both plants and human lungs. The two chemical compounds most often involved in air pollution are nitrogen oxides (NO_x) and sulfur dioxide (SO_2). Nitrogen oxides are emitted from power plants, factories, and motor vehicles, whereas 80 percent of the sulfur dioxide comes from coal-burning power plants. Once in the air, both of these gaseous substances are converted to tiny particles—nitrates and sulfates—that make up the annoying haze that reduces visibility. Plumes of acid-forming air pollutants originating in the Ohio Valley and other midwestern states from coal-burning electric power utilities and heavy industries are transported eastward by prevailing winds. When precipitation occurs, these particles reach the ground as acid rain or acid snow.

Great Smoky Mountains National Park experiences some of the highest air pollution of any national park. Although air quality in most urban areas throughout the country has been improving over the past two decades, air quality in the park had been showing signs of further deterioration. The most recent information (1994–2004), however, shows that air quality at the park is either improving (visibility and fine particles) or remaining stable (ozone concentrations and acid deposition). Air quality monitoring and research in the park shows that air pollution is impacting streams, soils, vegetation, visibility, and potentially public health. Threats posed by air pollution have placed Great Smoky Mountains National Park on the 2004 list of America's Ten Most Endangered National Parks. This list, which has been distributed annually since 1999 by the National Parks Conservation Association (NPCA), marks the park's sixth consecutive appearance on the list. Named in 2002 as the most polluted national park in the country, poor air quality in the Smokies often rivals urban areas. Clouds hanging over sensitive spruce-fir forests at Clingmans Dome and other high elevation sites are often as acidic as vinegar. In 2002 the park completed its first year of monitoring for toxic mercury and found the Smokies has some of the worst mercury pollution among monitored sites. Power plants emit nearly 40 percent of the mercury air pollution in the United States.

The spectacular overlooks for which this park is known are severely impaired by regional haze. Sulfates are the primary cause of visibility impairment. Under natural conditions, views extended for more than 100 miles, but because of air pollution, park visitors can expect to see only 25 miles on average. Occasionally, however, spectacular views are possible. One such day occurred on October 21, 2006, when my family and I hiked to the Clingmans Dome tower and could clearly see Mount Mitchell 73 miles to the northeast. During the summer months, visibility drops to an average of approximately 14 miles, making the Smokies one of America's haziest parks.

Carbon can make up 40 percent of the fine mass and contribute 20 percent of the haze in the park. Determining the exact origin of all of the carbon is still an unanswered scientific question. Some of the possible carbon sources include fresh, mobile source, urban emissions, aged urban carbon, secondary organics from trees, agricultural burning, and secondary organics from petrochemical sources.

Air pollution is causing damage to the park's vegetation. The Smokies has some of the highest ozone exposures at levels harmful to plants of any eastern national park. Ozone is formed from the oxides of nitrogen and hydrocarbons, both automobile exhaust gases, in sunlight. It is worse in high-elevation areas. Currently 90 plant species in the park, including black cherry, tuliptree (yellow poplar), sassafras, and sweetgum trees along with milkweed, blackberry, and cutleaf coneflower exhibit ozone-like foliar injury caused by the reaction of sunlight with pollution in the air.

Among the symptoms are purple, brown, or black spots on the top surface of the leaves of older foliage and between the veins of leaves. Acid rain, the result of sulfur and nitrogen oxides released from burning fuels, is suspected of causing damage to the red spruce trees at higher elevations, in addition to reducing visibility by 60 percent. Chronic and episodic acidification is adversely affecting high-elevation-sensitive streams and soils. Wet nitrate deposition had been increasing (1981–2003) and wet sulfate deposition had been decreasing at the park (1981–2003), but the trend in nitrate and sulfate concentrations in precipitation has been stable over the past decade (1994–2004). The nitrates deposited at the park are six to seven times the amount that local soils can naturally process, threatening sensitive plants and aquatic life.

Air quality monitoring station at Look Rock along the western edge of the park.

The park currently has seven air quality monitoring stations (Look Rock, Cades Cove, Cove Mountain, Elkmont, Clingmans Dome, Noland Divide, and the Purchase) that monitor levels of air pollution continuously. The Clingmans Dome air quality monitoring station was established in 1993. In 2005 cloud and dry deposition monitoring equipment was also installed at this site. The Smokies is one of only two sites in the eastern United States that has an automated cloud collector. (The other site is on Whiteface Mountain, New York.) Acid and mercury deposition are monitored at the Clingmans Dome and Elkmont sites. The park's air quality monitoring program collected nearly 10 million measurements during the 2006 field season. Data show that cloud deposition at Clingmans Dome accounts for over half of the total acid deposition (wet, dry, cloud), averages 3.6 pH (100 times more acidic than natural rainfall of 5.6 pH), and shows that the site is in clouds approximately 50 percent of the time. The park's ozone monitors recorded ozone exceedances of the eight-hour standards to protect public health on 217 days between 1998 and 2007: 42 days in 1998, 52 days in 1999, 31 days in 2000, 14 days in 2001, 42 days in 2002, 10 days in 2003, 3 days in 2004, 9 days in 2005, 9 days in 2006, and 5 days in 2007 (as of July 11, 2007). On April 15, 2004, EPA designated the entire park as nonattainment for the eight-hour National Ambient Air Quality Standard.

Research is currently under way in the park to study the effects of ground-level ozone pollution on "bio-indicator" species of native wildflowers and trees. Researchers are studying a number of factors including visible foliar injury, growth, photosynthesis, genetics, drought interaction, ozone exposures, and microclimate in determining the ozone sensitivity of a variety of species such as cutleaf coneflower, milkweed, black cherry, and yellow poplar. Biomonitoring gardens were established during the summer of 2001 at Twin Creeks, Great Smoky Mountains Institute at Tremont, and at the Purchase. Several sensitive species were planted, grown, and are being studied at these sites. Having two low-elevation sites and one

Blackberry leaves showing ozone damage (60 percent stippling).

high-elevation site will help the park understand how different ozone exposures affect these plants.

One study of ozone exposure measurements showed concentrations decreased as one descended into the canopy from above. Concentrations near the ground were about half those measured one meter above the canopy. This work is important to provide baseline data and to show trends in air pollution as reductions in sulfur dioxide and nitrogen dioxides are made over the course of the next decade.

The park has two Webcam sites that display real-time views, visual range, ozone concentrations, particle pollution, and weather conditions. One was installed at the Purchase Knob Appalachian Highlands Science Learning Center in August 2003. It averaged over 26,000 hits each week during 2004 with a weekly high of over 48,000 hits. As of February 2007 this site ranked seventh on the 10 most accessed Web pages monitored by the National Park Service. The Web site address is http://www2.nature.nps.gov/air/Webcams/parks/parks/grsm/grsmpkcam/grsmpkcam.cfm.

The other air quality Webcam site is at Look Rock north of Cades Cove. It receives between 40,000 and 60,000 hits per week. The web address is http://www2.nature.nps.gov/airWebcams/parks/parks/grsm/grsmcam/grsmcam.cfm.

As of 2004 air quality across most of the important areas of concern at Great Smoky Mountains National Park is improving. Fine particle concentrations, visibility on the haziest days, ozone concentrations, and acid rain have all significantly improved over the past five years. The park's air quality still has a long way to go in restoring and remedying the problems, but it is encouraging to see these improvements in conjunction with emission reductions.

The following Web sites can provide access to air quality data:

> Ozone and Weather: http://12.45.109.6/pls/portal30/get_input.show_parms
> Visibility and Fine Particles: http://nadp.sws.uiuc.edu/
> Acid Precipitation: http://vista.cira.colostate.edu/improve/
> Dry Deposition: http://www/ epa.gov/castnet/data.html
> Mercury: http://nadp.sws.uiuc.edu/mdn/
> UV radiation: http://oz.physast.uga.edu/introduction.html
> Forecasting and Mapping: http://www/epa.gov/aimow/ozone.html

Water Quality

Rain that has not been affected by human activity has a pH of approximately 5.6. The pH of pure rain is due to the presence of carbon dioxide, which reacts with water, producing carbonic acid that dissociates into hydrogen ions and bicarbonate ions. Thus, acid rain is any precipitation with a pH below 5.6. Rainfall in the park normally ranges between pH 4 and pH 5.

Water quality monitoring has been taking place for a number of years in the park. For the most part, park waters remain mostly free of chemical pollutants.

However, threats from human activities outside the park, from the 9,416,734 park visitors in 2006 (a 2.4 percent increase over 2005), and from nature all play a role in affecting the quality of the park's waters. As mentioned earlier, acidic deposition, combined with the Smokies' acidic bedrock, threatens aquatic ecosystems. In some places, nitrogen levels in the water have reached dangerous levels. Borderline environments may deteriorate to inhospitable conditions for some species. Storm event monitoring in several streams consistently shows a one- to two-unit drop in pH during storms, with the pH dropping below 5.0 in some streams and in one case down to 4.0. At soil pH of less than 4.5, nutrients such as calcium, potassium, and magnesium are leached and root-toxic aluminum is mobilized. Long-term monitoring of base flow stream water quality shows a trend of slowly declining pH at elevations below 4,000 feet. If current trends continue, all GSM waters in this elevation range will be below pH 6.0 within 14 to 51 years and eventually below 5.0 depending on elevation.

All water in the park needs to be boiled, filtered, or treated by some other acceptable method prior to drinking because of the possible presence of a protozoan parasite known as *Giardia lamblia*. Giardiasis occurs worldwide. It is found not only in humans, but in many other mammals including dogs, horses, and boars. The disease is generally characterized by chronic diarrhea that usually lasts one or more weeks. As with most other protozoa inhabiting the intestinal tract, the life cycle of *Giardia* involves two stages: trophozoite and cyst. Trophozoites stay in the upper small-intestinal tract, where they actively feed and reproduce. When the trophozoites pass down the bowel, they change into the inactive cyst stage by rounding up and developing a thick exterior wall, which protects the parasite after it is passed in the feces. People become infected indirectly by drinking feces-contaminated water. After the cyst is swallowed, the trophozoite is liberated through the action of digestive enzymes and stomach acids and becomes established in the small intestine. If stream water must be used for drinking, it should be boiled for one minute to kill *Giardia* as well as other infectious organisms that might be present. Chemical disinfectants such as laundry bleach or tincture of iodine may also be used to disinfect water of uncertain purity. These products work well against most bacterial and viral organisms, but are not considered as reliable as heat in killing *Giardia*. If water is cloudy, it should be strained through a clean cloth to remove any sediment or floating matter. Then the water should be boiled or treated with chemicals.

Native Pest Species

Southern Pine Beetle

Beginning in the fall of 1999 and continuing through most of 2000, the native southern pine beetle (*Dendroctonus frontalis*) killed patches of pine and red spruce throughout the park. Such outbreaks are cyclic and are controlled by native preda-

An example of southern pine beetle damage.

tors and parasites as well as weather patterns. Mild winters allow beetle populations to grow, and late-season dry conditions stress the trees, making them vulnerable to attack. The outbreak seemed to slow during 2001. Research in the park suggests that pine forests, southern pine beetles, and fire form an ancient triangle of interaction, in which large numbers of beetles dramatically increase dry, resinous fuels in spots on south-facing slopes. Fire burns more intensely there, creating the mineral soil "seedbed preparation" required for pine germination. In the absence of fire, however, pine beetle activity may lead to the rapid loss of pines, especially table-mountain pine, from the canopy.

Exotic Pest Species

Would you be startled to see a two-foot-tall monkey crossing the road near the Sugarlands Visitor Center? Two have been captured in the park and returned to their owners. What would you think about rounding a bend on Low Gap or Snake Den Trail near the Cosby campground and coming face-to-face with a 600-pound Texas Longhorn cow? Such incidents occurred in January 2003 before the cow was captured and reunited with its owner in Jones Cove. A feral ferret was captured in 2000 near the Forge Creek turnaround. A three-and-a-half-foot-long iguana was captured in the Cosby campground in September 2001. In February 2003, there were reports of an emu-like bird along the Appalachian Trail between Mount Collins and Double Springs shelters. Peacocks were reported near Mount LeConte, and an anteater has reportedly been seen in the park. Each of these represents a one-time occurrence and is the result of either an animal that has escaped from captivity or one that has been intentionally released into the national park because it is thought that it could survive in a wildlife sanctuary. Although these represent exotic species to the park's fauna, they are not damaging to other species nor to specific ecosystems.

Exotic species are species that have never been part of the original ecosystem. They have evolved and inhabited other regions of the world until being transported, usually through human activities, to regions of the world where they never occurred. Most exotics blend well with native species, but some create problems. Many are small organisms such as the balsam woolly adelgid, although one of the park's worst exotics is the European wild boar. Non-native plants, species that have been introduced to an ecosystem by seeds that are blown in by wind or carried by birds, vehicles, and humans are a threat to many of the park's ecosystems. Of the approximately 400 non-native species in the park, 60 spread aggressively, out-competing native plants for habitat (see Exotic Plants).

Balsam Woolly Adelgid

The balsam woolly adelgid (*Adelges piceae*) is a small, wax-covered insect pest that infests and kills stands of Fraser fir in the spruce-fir zone. This fir occurs naturally only in the southern Appalachians and used to be the dominant tree at the highest elevations. The adelgid was unintentionally introduced on trees imported from Europe in 1908, and the fir has little natural defense against it. The adelgids, which occur in great numbers, cover themselves with masses of "woolly" wax. Waxy patches on the stems and branches of the fir first announce the adelgid's presence. Later, the skeletons of dead trees are striking evidence. The adelgids attach themselves to the tree with their long, sucking mouthparts. The wax they exude protects

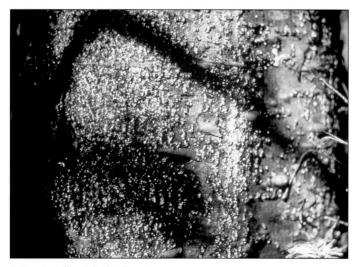

Balsam woolly adelgid infestation.

them from weather and predators. The adelgid feeds through cracks in the bark of the mature trees, injecting the tree with saliva containing toxins that cause abnormal cell growth in the tree. The firs overreact to the feeding adelgids, clogging their transport tissues, and literally starve to death. Trees usually die within five years of infection. Young trees have relatively smooth bark and are not attacked by the adelgid until they are large enough to produce roughened bark—approximately 20 years—which is also about the time they start to produce cones.

On the average, each female lays about 100 eggs. The eggs hatch into "crawlers" which move away from their parent. They search out a new location, attach, and begin to feed. Through a series of stages, adulthood is reached.

The adelgid was first recorded in the Great Smoky Mountains in 1963. In a little over 40 years, an estimated 70 to 90 percent of the mature Fraser firs in the park have been killed or are dying. Thousands of dead snags are all that are left of the mature trees on the highest mountain peaks. Fraser fir has been listed as a candidate species of special concern, and the spruce-fir ecosystem is also at risk.

The park service is trying to save some trees by spraying them with an insecticidal soap solution, but this can save only a few trees. Adelgid populations have been especially high in some areas such as the Balsam Mountain Road area for several years. Two treatments with insecticidal soap, which acts as a desiccant, were applied to the trees in this area each year from 1997 to 2000. A big drop in adelgid populations was evident in 2001 and only one treatment was needed. Treatment has continued each summer through 2007.

Ground surveys of the high elevation peaks in 2001 found that some mature Fraser fir were surviving despite over 35 years of balsam woolly adelgid infestation. Ten stands of mature fir over one hectare in size were documented. These surviving stands, as well as large individual trees and smaller stands, are being studied to determine their mechanisms of resistance. Researchers at North Carolina State University have cooperated with the park in studies to determine mechanisms of resistance in bark chemistry. Some firs have elevated levels of plant hormones such as juvabione that may prevent BWA from completing its life cycle. Other possible factors include bark thickness, past logging history, and microclimates.

The surviving Fraser fir on high-elevation peaks remained in relatively good health during 2005. Adelgid densities on monitor trees at Clingmans Dome, Mount LeConte, and Mount Sterling remained low. Balsam Mountain monitor trees had lower adelgid densities than in 2004, but still required insecticidal soap treatment. With the exception of Mount Sterling, nine of the ten study sites showed vigorous new growth and little to no adelgids present in 2005. Fir regeneration in all of these areas as well as the Mount Guyot sites is growing at a rapid rate, leaving great hope for survival of future stands. Mount Guyot continues to have the largest, healthiest remaining stands of mature fir in the park.

Research continues on trying to find methods to control the adelgid. Seeds have been collected and preserved so the genetic strain will not be lost. Seedlings are grown for planting at a location in the park (Purchase Knob) where the trees can be treated. In 1995, 600 Fraser fir seedlings, grown from seeds collected in the park, were planted. While some have become infested with BWA, they are monitored annually and, if needed, treated with insecticidal soap or a dormant oil spray mix. The plantation is a cooperative effort between the park and the University of Tennessee Department of Forestry, Wildlife and Fisheries. The planted trees are managed as a reservoir of the park's Fraser fir genetic material.

The loss of Fraser fir trees has a detrimental effect on the flora and fauna of the habitat. Currently, populations of eight species of mosses and liverworts that live in association with the firs are declining. Vertebrates such as the pigmy salamander and the northern flying squirrel are among other vulnerable species.

Chestnut Blight

The American chestnut tree was an essential component of the entire eastern United States ecosystem. Approximately 25 percent of eastern woodlands from Maine to northern Georgia and from the Piedmont west to the Ohio Valley (over 200 million acres) was composed of chestnuts. It was thought to have represented roughly 40 to 45 percent of canopy trees in some pre-blight southern Appalachian forests. The American chestnut was as plentiful as oaks and maples are today. In those thousands of years preceding the blight's arrival, an enormously complex set of relationships evolved which tied the chestnut to innumerable bird, mammal, and insect species and other organisms, as well as to rocks, water, soils, and fire. A late-flowering and productive tree, it was unaffected by seasonal frosts, and the nuts had a high fat content, making it the single most important food source for a wide variety of wildlife from the white-footed mouse to black bears and turkeys. Chestnuts at one time produced about 50 percent of the entire forest nut crop. These were big trees with some attaining trunk diameters of nine to 10 feet and reaching heights of 120 feet. Some survived for 400 years. Foresters regarded the American chestnut as the best hardwood timber tree in America. Known as the "redwood of the East," its lumber was straight-grained, easily worked, exceptionally durable, and of the highest quality. It was easily split into fence rails and shingles. Since the wood was highly resistant to the attacks of wood-destroying fungi, it was extensively used for fence posts, railroad ties, telephone and telegraph poles, and mine timbers. The fast-growing American chestnut was once the primary tree throughout much of the eastern forests, including the Great Smoky Mountains. It constituted between 20 and 30 percent of all of the trees in the park.

A parasitic fungus (*Cryphonectria parasitica*) was accidentally introduced into the United States in the late 1800s presumably on Asian chestnut trees. It was not discovered until 1904 in New York City and quickly spread throughout the eastern forests. The fungus is dispersed by spores in the air, on raindrops, or by animals. It is a wound pathogen, entering through a fresh injury in the tree's bark. It spreads into the bark and underlying vascular cambium and wood, killing these tissues as

American chestnut in bloom.

it advances. The flow of nutrients is eventually choked off to and from sections of the tree above the infection, killing them. Like many other pest introductions, it quickly spread into its new—and defenseless—host population. American chestnut trees had evolved in the absence of chestnut blight, and our native species lacked the genetic material to protect them from the fungus. By the 1950s, 4 billion American chestnut trees, about 99.9 percent of the Eastern population, succumbed to the disease. Most of the chestnut trees in the park were killed by this fungus during the 1920s and 1930s. Many of the fallen trees can still be seen along park trails. Even though the trees have died, the lethal fungus survives on decaying logs and leaves. Because the blight does not affect the chestnut's root system, trees still persist through re-sprouting. The sprouts grow for a few years before the blight girdles their trunks and kills them. Some sprouts grow large enough to bear

American chestnut blossom (close-up).

American chestnut burrs.

a few chestnuts before they die. During the summer of 2003, Matthew Wood, a forestry technician for the park, hiked hundreds of miles in the Smokies looking for American chestnut trees that had reached flowering age without falling prey to the chestnut blight. Overall, he verified the location of more than 200 survivors, the largest being a tree near Gregory Bald that measured one foot in diameter and stood roughly 55 feet tall. He found flowering chestnuts from Andrews Bald (5,800 feet) down to about 2,000 feet elevation.

Chinese chestnut trees are resistant to the blight. Whereas blighted North American chestnut trees die, blighted Chinese chestnuts suffer only cosmetic damage. But the Chinese chestnut lacks many of the characteristics of the American species. Most obvious is stature: the Chinese species is low-growing and spreading,

The chestnut blight killed most of the American chestnut trees in the park during the 1920s and 1930s.

much like an old apple tree, whereas an American chestnut can grow straight and strong to a hundred feet or more. This habit of growth combined with the quality of wood made the American chestnut a dominant forest tree species.

The American Chestnut Foundation (TACF) was founded in 1983 by a group of prominent plant scientists who recognized the severe impact the demise of the American chestnut tree imposed upon the local economy of rural communities and upon the ecology of forests within the tree's native range. The foundation is using the backcross method of plant breeding to transfer the blight resistance of the Chinese chestnut to the American chestnut. Backcrossing is the standard method for transferring a single trait into an otherwise acceptable plant. For chestnuts, it entails crossing the Chinese and American trees to obtain a hybrid which is one-half American and one-half Chinese. The hybrid is backcrossed to another American chestnut to obtain a tree which is three-fourths American and one-fourth Chinese on average. Each further cycle of backcrossing reduces the Chinese fraction by a factor of one-half. The idea is to dilute out all of the Chinese characteristics except for blight resistance. The goal is to produce trees which are fifteen-sixteenths American, one-sixteenth Chinese. Many of these hybrid trees will be indistinguishable by experts from pure American chestnut trees. The first line of blight-resistant American chestnuts is expected to be ready for planting by about 2008. Researchers hope that this effort will become the most successful nature restoration program in the nation's history.

Dutch Elm Disease/Elm Yellows

Both of these diseases have been found in the park. Dutch elm disease (DED) is a fungal disease that originated in Europe and has killed many elms in America. It was introduced into this country about 1930. Elms become especially susceptible when beetles carrying the fungus reach the canopy and begin feeding. Dying elms were first noticed along Little River starting in the late 1980s. Elm yellows is a mycoplasma-like organism (MLO) disease transmitted by leafhoppers. Visible symptoms of the disease include yellowing of foliage in summer. Once infected, elms do not recover.

Dogwood Anthracnose

Flowering dogwood trees are being infected by *Discula destructiva*, a destructive fungus from the Orient that causes dogwood anthracnose. It was first identified in the United States at Chehalis, Washington, in 1977, and shortly thereafter in the New York City area. It spread southward and was found in Maryland in 1983 and in north Georgia in late 1987. Widespread dogwood mortality has resulted in its path. It began affecting the park in the late 1980s. Symptoms of anthracnose are not evident until leaves are fully expanded, but a cool, wet spring provides good conditions for the fungal disease. Loss of the dogwood's floral displays and very high protein-rich berries for birds and other wildlife is significant. What could

Dogwood leaves contain the highest concentration of calcium of any eastern deciduous tree. The loss of dogwood trees by a fungus that causes dogwood anthracnose could have far-reaching effects.

be most disruptive to natural processes in the park, however, is the loss of the element calcium in forest soils resulting from dogwood mortality. Dogwood foliage, which contains the highest concentration of calcium of any eastern deciduous tree (approximately 3 percent) and has a rapid decomposition rate (64 percent reduction in litter mass after two years), is a prime soil builder. They may very well act as a "pump" of calcium as they drop their leaves on the forest floor each year, and calcium is released through decomposition. Calcium is a critical ingredient for cell wall development in all plant species. Understanding calcium loss at higher elevations is a high priority due to very large depositions annually of acid precipitation.

Heavy mortality of dogwood has occurred in the park since the late 1980s, ranging from 57 percent in oak-pine forests to 94 percent in acid cove forests. Mortality was highest among smaller trees, suggesting that as larger trees die and are not replaced by new trees, the species will largely disappear from the park. As dogwood trees die, they have largely been replaced in the understory by eastern hemlock, a species that has greatly increased in importance as a result of fire suppression and loss of American chestnut (chapter 5). Changes in soil chemical properties may be amplified by the increased importance of hemlock since soils under hemlock trees are typically more acidic. Consequently, replacement of hardwood species by hemlock may result in lower soil pH and reduced availability of many nutrients. Eastern hemlocks, however, are currently under attack by a non-native adelgid (see page 225). One can only speculate about the forest composition if a majority of the hemlocks disappear.

An area burned by a wildfire in 1976 was found to have a 200 percent increase in dogwood density. Since anthracnose favors cool, damp habitats, it is possible that the sunny, drier conditions that exist in open areas after fire may inhibit its spread. A study completed in 2006 that examined the response to fire of flowering dogwood in oak-hickory stands compared dogwood density, dogwood foliar health and crown health, stand structure, eastern hemlock density, plot species richness, and plot diversity among four sampling categories: unburned stands, and stands that had burned once, twice, and three times over a 20-year period (late 1960s to late 1980s). Burned stands, especially double-burn stands, had significantly greater flowering dogwood stem densities than unburned stands. In addition, the density of eastern hemlock, a species that creates stand conditions favorable for dogwood anthracnose, was greatly reduced in burned stands. Past burning did not drastically affect overall overstory or understory species composition. Total overstory density was greater in unburned stands, but understory stem density was greater in burned stands than unburned stands. The results of this study indicate that prescribed burning may offer an effective and practical management tool to reduce the impacts of dogwood anthracnose and prevent the loss of flowering dogwood in oak-hickory forests.

Since dogwood foliage is an important source of calcium for the rich biota of the forest floor, changes in understory composition in conjunction with the widespread loss of dogwood trees may greatly impact numerous ecological relationships in these forests, including calcium availability, nutrient cycling, and food source availability for wildlife. Several recent studies have examined the significance of calcium to various tree species and to the survivability of dogwood trees infected with anthracnose. One study found significantly greater levels of calcium in high-density dogwood plots than in low-density plots in both cove hardwood and oak hardwood forest types. In both forest types, calcium mineralization occurred primarily in the forest floor and not in the mineral soil. These results indicate that the loss of dogwoods due to dogwood anthracnose has altered the calcium cycle and may negatively affect the health of eastern hardwood forests. A study begun in 2004 examined the contribution of three soil cations (potassium, magnesium, and calcium) to dogwood survival following anthracnose infection. While most seedlings died after one season of exposure to dogwood anthracnose, seedlings that had lower inputs of calcium and potassium showed higher levels of disease severity sooner than seedlings in other treatments, suggesting that these nutrients play a role in dogwood's survival from anthracnose. Magnesium levels did not appear to have an effect on disease severity or mortality. Another study showed that at a given level of soil calcium availability, dogwood trees contained greater concentrations of calcium than three other dominant understory species (eastern hemlock, red maple, and rosebay rhododendron). Between 1977 and 1979 (pre-anthracnose) and 1995 and 2000 (post-anthracnose), the annual calcium contributions of understory woody vegetation declined across all forest types, ranging from 26 percent in oak-pine stands to

49 percent in acid coves. In oak-hickory and oak-pine stands, large increases were observed in the foliar biomass of eastern hemlock, a species whose calcium-poor foliage increases soil acidity. Calcium cycling in oak-hickory stands was more negatively affected by the loss of dogwood trees than the other forest types.

Although the long-term effect of this disease in the park is not fully understood, the results of studies are not encouraging. Research shows that the last dogwoods to succumb in a site are the ones in sunny locations. These may survive years after nearby shaded trees are dead. Trees are dying in most watersheds along streams, on northerly slopes and the cooler, moister high elevations of the park. There is no treatment known to be practical, affordable, or environmentally advisable for use in the park.

Gypsy Moth

In 1869, gypsy moth larvae from France that were being evaluated by Leopold Trouvelot for silk production were blown from a window sill in Medford, Massachusetts. The first outbreak of European gypsy moth (*Lymantria dispar*) occurred about 10 years later, and by 1889 it was doing heavy damage in certain parts of the Boston area. By 1987 the gypsy moth had established itself throughout the Northeast. It is one of the most damaging pests of hardwood forests and urban landscapes, defoliating a million or more forested acres annually. Although capable of feeding on over 300 species of trees and shrubs, the moths prefer oaks. Gypsy moths are spread one of two different ways. Natural spread over short distances occurs as newly hatched larvae spin short lengths of silken thread which allow them to be blown by the wind. Over the last 10 to 15 years, gypsy moths have moved long distances on outdoor household articles such as cars and recreational vehicles, tents, firewood, household goods, and other personal possessions. An estimated 85 percent of new infestations have been through the movement of outdoor household articles.

The gypsy moth has four different life stages: egg, larva or caterpillar, pupa, and adult moth. Female moths lay eggs in sheltered areas. Each egg mass contains

Male gypsy moth.

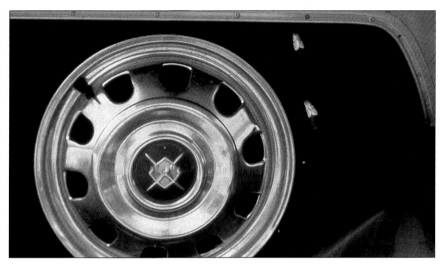

The spread of gypsy moths has been inadvertently aided by human activities and movements. Eggs deposited on tires can be transported a considerable distance.

between 500 and 1,000 eggs and has a tan, fuzzy appearance. The egg masses can be found all winter long but will not hatch until spring. Hatching usually coincides with the budding of most hardwood trees. Only the caterpillar stage of the gypsy moth feeds. When fully grown, the caterpillar will be approximately two inches long, very hairy, and have five pairs of blue dots followed by six pairs of red dots along its back. The larval stage lasts about seven weeks. Gypsy moth pupae are about two inches long, dark brown, and are lightly covered with hairs. Pupation usually occurs in protected areas of the tree. Adult moths have a distinctive

Traps baited with pheromones from female gypsy moths are used to attract male moths and assess population density.

Gypsy moth larvae eating leaves. Only the caterpillar stage of the gypsy moth feeds.

inverted V-shape that points to a dot marking on their wings. The dark brown males are smaller than the females, and have feathery antennae. Female moths have creamy white wings with a tan body. Although the female possesses wings, she is unable to fly. (Both males and females of the recently introduced Asian gypsy moth which was discovered near Wilmington, North Carolina, in June 1993 can fly. This dramatically increases the likelihood of spread.)

Gypsy moths are currently near the Virginia-Tennessee border and threaten oak forests with total defoliation. Trapping in the park using pheromone traps to lure male moths was begun in the late 1980s by both the park Resource Management staff in park campgrounds and picnic areas and the Tennessee Division of Forestry. Several male moths were caught in traps at Metcalf Bottoms and Elkmont. Trapping has continued in the park each year with between 129 and 160 traps in operation. No additional gypsy moths have been taken in the park, although the Tennessee Department of Agriculture has had positive traps in Sevier County for several years and has conducted small-scale spraying operations using the broad-spectrum insecticide diflubenzuro which acts as a chitin inhibitor. Sevier County is one of three Tennessee counties that is classified as a "known infested county," where more than one life stage has been found.

Beech Bark Scale Insect/Beech Bark Disease

This European insect (*Cryptococcus fagisuga*) and the *Nectria* fungus, which together cause beech bark disease (BBD), were first recorded in the park in 1993. BBD has now killed high-elevation beech forests throughout the park, and the disease has moved to individual trees at lower elevations. Ten long-term monitoring plots were established throughout the park in 1994. The plots are evaluated every other year. By 2000, the most severely affected plots had significant mortality of mature beech. One area (Fork Ridge) had dense resprouting from roots, while another (Forney Ridge) had very little beech regeneration. Time will tell how the disease will affect the resprouts. During 2003 beech trees continued to decline with heaviest mortality along the Appalachian Trail. During 2004 evaluations showed increased severity of the disease. One plot moderately infected in 1994 had 100 percent mortality of the overstory beech (37 trees). By 2005 the western Appalachian Trail section showed somewhat less severity of the disease, with surviving overstory trees showing evidence of scale attack and understory beech regeneration. The eastern portion of the park had significant beech mortality with areas of understory regeneration. Results of plot monitoring in 2006 revealed mortality of mature trees ranging from 20 percent to 100 percent.

Hemlock Woolly Adelgid

The hemlock woolly adelgid (*Adeleges tsugae*) is a tiny, soft-bodied, exotic invasive species that gets its name from its woolly white appearance and because its host is the hemlock tree. The adelgid has a complex life cycle and produces two generations per year. Eggs are brownish-orange and wrapped in a white fluffy substance secreted by an adult female. Reddish-brown nymphs (or crawlers) hatch from the eggs and use their threadlike mouthparts to pierce a hemlock branch and suck sap from the branch. These nymphs go through four stages before becoming adults and also wrap themselves with a white, fuzzy covering at the base of needles on the

The hemlock woolly adelgid is increasingly infecting the more than 74,000 acres of hemlock trees in the park. Non-native predatory beetles are being released as a biocontrol. Photo by Steve Bohleber.

underside of hemlock branches. Adults are reddish-purple with some having two pairs of wings. The flying adults leave the hemlock in search of a secondary oriental spruce host (which does not occur in the United States). The wingless adults stay on the hemlock host and produce 50 to 300 eggs. Adults, as well as the nymphs, feed at the base of hemlock needles, disrupting nutrient flow and eventually causing the tree to starve to death. The hemlock needles dry out and drop from the tree. This defoliation generally causes an infested tree to die within three to 10 years.

The hemlock woolly adelgid is native to Asia, where it is not a problem to native hemlocks. It was introduced into the United States in the 1920s in the Pacific Northwest, and in the early 1950s to the Washington, D.C., and Richmond, Virginia, areas. It lacks natural enemies in North America, so it has spread throughout the eastern United States via wind, birds, mammals, human activities, and the transport of infected nursery stock, creating an extreme amount of damage to natural stands of hemlock, specifically eastern hemlock (*Tsuga canadensis*) and Carolina hemlock (*Tsuga caroliniana*). The eastern hemlock is the only native hemlock in the park. The Carolina hemlock has been planted in Gatlinburg and occurs in northeastern Tennessee and in portions of Virginia, North Carolina, and Georgia, but it has never been recorded from within the park. The predicted spread rate of the adelgid is about 20 miles per year. Scientists believe the adelgid poses the greatest threat to the ecology of the Smokies since the chestnut blight devastated the region in the mid-1900s.

In 2001 the closest known site for the hemlock woolly adelgid was less than 10 air miles from the park's boundary. It was first discovered in the park in 2002 along Stony Branch two miles west of Cades Cove and has the potential to devastate hemlock stands in all age classes. It has nearly wiped out the hemlock forests further north in the Appalachians at Shenandoah National Park and elsewhere. Hemlocks are very important ecologically, with a number of other species dependent on them, so their loss would have a ripple effect throughout the park. In ravines, dense stands of the evergreens shade streams, keeping them cool in summer. This cooling effect (as much as 4° C) makes the streams hospitable for native brook trout and a host of other fishes and aquatic life. The shelter of hemlock groves is also a preferred habitat for several species of migratory songbirds including the blackburnian warbler, black-throated green warbler, wood thrush, and blue-headed vireo. George Farnsworth and Ted Simons found 84 percent of wood thrush nests in hemlocks in the park.

The Smokies contains the largest expanse of old-growth hemlocks in the East, approximately 800 acres, with some trees being six feet across, over 160 feet tall, and over 500 years old. As of 2006 a total of 21 hemlock trees over 160 feet in height have been documented in the park. Unfortunately, four of these have died. The park's total hemlock resource has been mapped at more than 74,000 acres with over 18,000 acres of hemlock-dominated forests. This species is ubiquitous throughout the low- to mid-elevation forests.

One striking example of the significance of this tree occurs during the winter months when most deciduous trees have lost their leaves. Other than an occasional pine tree, a rhododendron, or a small holly, the *only* green tree that one observes in driving the two miles from Gatlinburg to the Sugarlands Visitor Center is the eastern hemlock. If these trees were lost, what a tremendous difference it would make to the visual landscape and to this green corridor through which millions of park visitors travel. Hemlocks also dominate the roadsides during the winter months as one begins to travel toward Newfound Gap on the transmountain road or as one begins to travel from Sugarlands toward Cades Cove. Along these routes, there is a greater abundance of rhododendrons, dog hobble, and pines, which help add color to the landscape during the winter, but the loss of the hemlocks would still be dramatic.

The hemlock woolly adelgid (HWA) control project began in the park in 2002 and has greatly expanded over the past several years. By 2004, the project consisted of a full-time coordinator and six forestry technicians. In 2003, 38 50-by-20-meter monitoring plots in forest stands containing various amounts of hemlocks were sampled. These plots will provide baseline data of success or lack of success as the park continues to battle the adelgid. In 2005, some areas showed stress and decline (Cades Cove and Cataloochee), while others remained relatively healthy (Cosby). During 2005, HWA surveys were conducted throughout the park, concentrating on the nearly 800 acres of old growth, 18,000 acres dominated by hemlock, heavily visited developed areas, and roads. Unfortunately, infestations were identified in all major watersheds of the park, although there are still areas where HWA has not been seen. Although large-scale hemlock mortality has not yet been observed in the park, scattered areas of mortality throughout the park are expected in the near future. If control measures are unsuccessful, the hemlock woolly adelgid could eliminate park hemlocks and destroy the entire forest type.

Integrated Pest Management for HWA includes surveys, pre- and post-treatment assessments, and chemical and biological controls. Chemical control activities include foliar treatments with a quickly degradable insecticidal soap and soil injection of the systemic insecticide Imidacloprid. Although the soap itself is relatively inexpensive, application is extremely labor-intensive. The foliage of each tree must be completely saturated with the soap twice each year, and aerial spraying is not effective. On December 12, 2006, the 50,000th tree (since 2002) was systemically treated near Albright Grove on Dunn Creek. Foliar treatments done on an annual or semi-annual basis have covered 950 acres since 2005. All of the developed areas in the park have received an initial treatment, as well as some backcountry campsites, the Albright Grove Loop Trail, Boogerman Loop Trail, Rainbow Falls Trail, and Trillium Gap Trail. Treating trees in developed areas helps ensure visitor safety and aesthetics, and reduces maintenance costs. Treatment for a fifteen-inch hemlock costs an estimated $19 and will protect a tree for three to five years.

Forest technician injecting insecticide into mature hemlock trees. Photo by Jeremy Lloyd.

Biological control is the most effective long-term method for fighting the adelgid in large wildland areas. Releases of a non-native predatory ladybird beetle (*Sasajiscymnus tsugae*), a species-specific predator of the adelgid, as a bio-control began in 2002. These predator beetles are related to common native ladybug beetles but are about one-tenth their size and solid black rather than the traditional black and red variety. This predator beetle was carefully studied before approval for release. It feeds only on adelgids, overwinters in forest leaf litter, and does not congregate. No one is likely to see them since they are about the size of a pinhead. Beetles are released in mass quantities of 2,500 to 5,000 beetles per site. The selection of sites for release is based on certain criteria that the sites have enough adelgids to allow the beetles to find their prey quickly, feed voraciously, and reproduce. The University of Tennessee started rearing beetles and supplying the park in 2004. Although it is too early to assess the overall success of this biocontrol, preliminary monitoring results are encouraging. Releases to date are shown in Table 11.1.

A comprehensive monitoring program was initiated in spring 2005 to evaluate treatments in terms of HWA population levels and tree health. Officials estimate that several million beetles need to be released in the park to bring the hemlock adelgid under control. The cost to purchase beetles from rearing facilities is approximately $1 per beetle.

In March 2006 a second predator beetle, *Laricobius nigrinus*, native to the Pacific Northwest, was released in the park for the first time. Research is currently under way at Virginia Tech in Blacksburg on a Japanese beetle closely related to

Laricobius nigrinus. The new beetles are able to survive at lower temperatures, are more voracious, and are quicker to develop than their American counterparts. In a perfect scenario, the park would like to eradicate the adelgids, but researchers believe the adelgids can be kept in check by the populations of the two beetles.

Twenty-four Eastern Hemlock Conservation Areas were established in 2006, ranging from 15 to 123 acres. All of the areas selected are old-growth hemlock forests and are distributed throughout the park at varying elevations. These areas receive a combination of systemic and biocontrol treatments.

Table 11.1. Predator beetle releases in the Great Smoky Mountains National Park.

Year	Releases	# of Beetles
2002	10	29,945
2003	7	21,546
2004	13	35,533
2005	36	77,083
	9 (sites)	40,514 (eggs)
2006	27	65,849
2002–2006	93	229,956

Pear Thrips

Thrips are tiny insects often found in flowers, leaf litter, and decomposing wood. In 2005, specimens of pear thrips were discovered at several locations in the park for the first time. They had been passively collected in Malaise (insect) traps during 2005. This species was introduced from Europe and feeds on a variety of fruit trees, as well as maple, basswood, birch, beech, ash, and black cherry. Recently, this species has been implicated in the widespread death of sugar maples in the northeastern United States.

Butternut Canker

Butternuts, also called white walnuts, are not numerous in the park; a foreign fungus called stem-canker fungus (*Siroccus clavigigenenti-jugulandacearum*), whose spores are spread by wind, rain, and insects, is making them less so. The butternut canker produces elongate, lens-shaped cankers on the trunk, limbs, and twigs and even penetrates the immature nut, killing the embryo and preventing reproduction. The fungus is believed to have been introduced into the United States from Asia around 1960 but went unnoticed in the southeastern United States until about 1986.

Seventy permanently marked butternut trees have been monitored in the park since 1987. They are evaluated every three years. In 2005 the health of these trees varied widely, but generally trees that received adequate sun have some branch dieback and few observable cankers, some of which have healed over with time.

In the 1990s the park and the University of Tennessee grafted new twigs (scions) to other rootstock and began screening these seedlings and those grown from collected nuts of the park's remaining butternuts before the parent trees die.

Fire Ants

Fire ants (*Solenopsis* sp.) were discovered in the park for the first time in 2002. They were found in several areas in Cades Cove, including along Hyatt Lane, at the riding stables, and along the Foothills Parkway near the Look Rock parking area. Fire ants were introduced into Mobile, Alabama, from southern South America in 1918.

Fire ants are small (1.6 to 6.3 mm long), reddish to brown ants that are very aggressive when their mound is disturbed. A mature fire ant colony can contain over 250,000 ants. Their effects on recreation, agriculture, and other outdoor activities are well known. When disturbed, they are quick to attack humans and other animals. They use their jaws to push up a fold of skin while stinging it repeatedly. Generally, a dozen or two are attacking at the same time. Their painful sting injects a venom which produces a burning sensation like fire, hence their common name. They often damage young plants and are pests of ground-nesting birds and wildlife which they sting to death and devour. They are expected to have a significant impact on certain aspects of the natural ecology of the park, especially in the drier habitats.

European Mountain Ash Sawfly

The European mountain ash sawfly reached Canada from Europe in the early 1900s. It was first observed in American mountain-ashes in high-elevation areas of the Smokies in the 1980s, killing half of the ashes in some plots. Defoliation prevents the production of food and depletes the tree's energy reserves. The red berries serve as food for ruffed grouse and other birds.

European Wild Hog

It is not known exactly how or when the wild hog came to the park. Thirteen young European wild boars, weighing 60 to 75 pounds apiece, first arrived in Murphy, North Carolina (approximately 40 miles south of the park), in April 1912, destined for a game preserve on Hooper Bald, near Robbinsville (about 12 miles south of the park) where they were released. The owner of the game preserve was George Gordon Moore, an American businessman. It is believed that the animals had been purchased through an agent in Berlin, who said they came from Russia. About 1920, an estimated 60 to 100 boars escaped from the preserve. As they dispersed, they hybridized with free-ranging domestic pigs and made their way into the Great Smoky Mountains National Park. It is believed that they entered the southwestern quadrant of the park near Calderwood in the late 1940s. The invasion steadily spread from west to east, averaging approximately 2.75 kilometers per year. They are currently found throughout the park, inhabiting a wide variety of habitats including forests and grass balds.

The body of a wild hog is built somewhat like that of a bison, being higher and heavier in the shoulder region. It is covered with thin, coarse hair. Hogs are usu-

The European wild hog, a non-native (exotic) member of the park's fauna.

ally black, but the tips of the guard hairs are silvery-gray or brown. A mane of long bristles may develop down the back. The upper tusks are distinctive in that they curve upward as they grow. Some individuals may stand three feet (900 mm) high at the shoulder and weigh over 400 pounds (180 kg), although most are considerably smaller. The largest hog that has been captured in the park has been a male weighing 303 pounds (136 kg) taken on Welch Branch in September 1962.

Hog depredations were first noted on Gregory Bald and along the state line in 1958 where overturned sod was observed, a result of their rooting. Some feel that this may have been a factor in the invasion of trees on the bald, for tree seeds landing on the unprotected soil probably had a better chance to germinate than if they had landed on the thick grass mat. In 1959, it was found that these animals were concentrated in the area between Cades Cove and Fontana Lake. Wild hog control began during August 1959, at which time there were an estimated 500 hogs in the park. Methods include trapping, shooting, and limited fencing. Historically, trapping was the primary method of control, but since 1993, 63 percent of the hogs have been removed by shooting. Between 1959 and 2005, a total of 10,392 wild hogs have been removed from the park, of which 5,672 (54.6 percent) were from the North Carolina side and 4,626 (44.5 percent) were from the Tennessee side. Location coordinates were not recorded for 85 animals (0.8 percent). Of these, 4,006 hogs have been donated to the North Carolina Wildlife Resources Commission (NCWRC) and to the Tennessee Wildlife Resources Agency (TWRA) to be relocated to state game lands. However, because of the concern of transporting hog-related diseases, the states of Tennessee and North Carolina terminated the relocation of wild hogs from the park. Several wetlands with unique habitats and rare species as well as a high-elevation beech gap have been fenced (exclosures) to exclude wild hogs.

Typical hog damage on a grass bald.

The population density of wild hogs in Great Smoky Mountains National Park is unknown but is thought to fluctuate drastically with available food resources. During years when the fall mast crop, especially acorns, is poor, overall reproduction is low. It also results in a low removal rate, particularly for younger animals. Since 1991, oak mast failures in the park have occurred in 1992, 1997, and 2003. The proportion of adults removed the years that followed were 78.0 percent, 78.6 percent, and 79.3 percent, respectively, compared with an average of 58.3 percent in other years.

On January 2, 2003, the wild hog control program hit a milestone when hog number 10,000 was removed from the park. Park wildlife managers continue to receive reliable reports of feral swine being released along the park boundary. These hogs have moved well inside the park boundary as evidenced by several hogs removed during 2003 that were spotted with physical characteristics more typical of a feral hog. European wild hogs are typically black in color. In 2006, 254 wild hogs were removed from the park.

Why are these mammals the most unwelcome animals ever to invade the park, and why are they referred to as the park's Public Enemy #1? Hogs feed on mast as well as invertebrates and small vertebrates. Each wild hog is a little bulldozer, rooting up vegetation in its search for food. During the spring and summer, most wild hogs move to the ridges at elevations above 4,000 feet where they feed and rear their young in the northern hardwood forests. On some of the grassy balds, large areas of bare soil have been exposed where the sod has been uprooted by hogs in their search for food. Roots and herbs, especially spring beauty corms, are important in spring and summer diets. Salamanders are especially common, with an average of 1.75 found per stomach at the higher elevations. The most common salamander taken is the endemic red-cheeked salamander. An investigation in 1979–1980 found that two mammals that depend largely on leaf litter for habitat, the redbacked vole and the short-tailed shrew, had been almost eliminated from intensely rooted areas of northern hardwood forest. Rooting accelerated the leaching of calcium, phosphorus, zinc, copper, and magnesium from the leaf litter and soil. Nitrate concentrations, however, were higher in soil, soil water, and stream water from the rooted stands, suggesting alterations in ecosystem nitrogen transformation processes. During the fall and winter months, hogs return to lower-elevation areas, where they feed primarily on acorns.

The wild hog has few predators. Bobcats prey on young pigs, and both young and adult hogs may be preyed upon by black bears and coyotes.

Each year, serum samples are collected from wild hogs and submitted to the North Carolina Department of Agriculture and Consumer Services (NCDACS) and the U.S. Department of Agriculture to be tested for swine brucellosis, pseudorabies, and hog cholera. These diseases are of concern to the local livestock industry as well as to public health agencies. This survey was initiated because in recent years several hogs removed from the park have appeared to have abnormal behavior characteristics and harbor physical characteristics (white or spotted coloring, curly tail, short snout) generally not found in this region. In fact, four animals removed from the park in February 2000 showed no fear of people, and two were completely white! The indiscriminate and illegal relocation of wild hogs raises immediate concerns regarding the potential transmission of animal diseases, primarily swine brucellosis, hog cholera, and pseudorabies.

From December 1998 to December 2004 a total of 348 wild hog blood serum samples were collected. All samples tested negative for these diseases. However, in 2005–06, serum samples were collected from 167 wild hogs, and three samples from North Carolina were seropositive for pseudorabies. Serological surveys have also revealed evidence of porcine parvovirus, leptospirosis, and toxoplasmosis.

Exotic Plants

Exotic plants are generally species that have been introduced to an ecosystem by human activities, although some may have been transported by wind, water, or animals. Seeds may be transported on vehicles and on the boots of hikers; seeds and/or plants may be in fill dirt used in construction projects. Exotics often have an advantage over native species because in their new environment there may be no natural controls such as parasites, predators, or diseases. Exotics are capable of causing major changes to an ecosystem and even threatening the survival of native plants and animals.

The park, with its many types of habitats, abundant rainfall, and relatively mild climate, provides suitable habitat for many species of plants, both native and exotic. Furthermore, since the climate and habitats in the Smokies closely resemble those in parts of eastern Europe and central Asia, plants from those areas have little trouble growing and reproducing in the Smokies.

As of August 2007 approximately 400 exotic species of plants have been identified in the park. Of these, some 60 species are aggressive and pose serious threats to the park's natural ecosystems. The park's vegetation management crew works to control invasive exotic plants by foliar spraying, cutting/spraying, annual cutting, pulling, and monitoring for after-action progress. As of August 2006 the exotic plant database contained over 900 active and inactive exotic plant sites. Among the species of greatest concern are kudzu, Japanese grass, privet, multiflora rose, Japanese honeysuckle, mimosa, garlic mustard, oriental bittersweet, musk thistle, English ivy, periwinkle, plume grass, orange-red hawkweed, Johnson grass, climbing euonymous, Norway spruce, Japanese spiraea, white poplar, Chinese yam, Japanese knotweed, bush honeysuckle, mullein, and tree of heaven.

Nature has a remarkable ability to adapt to changing conditions. As the chestnuts were lost, there were increases in other species of trees. Thus, the forest is continually adapting. The forest that our grandchildren and their descendants will see may be quite different in composition from the forest of today, but there will always be a forest in the Smokies.

CHAPTER 12

WHAT THE FUTURE MAY HOLD

Life for people is short on Earth, and we should not live selfishly only for the moment to suit our own needs. We must keep the future in mind. Learning from mistakes is only good if these mistakes can be avoided in the future, not if the damage done is beyond repair.

—Nan Fariss (2006)

The Great Smoky Mountains have changed many times in their long history, and no doubt they will change many more. Parts have been under the sea and have bulged up again into a mighty mountain range, only to wear down while trees covered their slopes and life continued to evolve.

Even in the last 70 years, the Smokies landscape has changed markedly. The many open fields and pastures have been reclaimed by forest, forcing animal life to adjust accordingly. This has diminished the presence of open-country animals such as red foxes, woodchucks, bobwhite quail, and rabbits, and encouraged black bears, pileated woodpeckers, and other forest animals.

What changes can we expect in the next 70 years? Within the park, probably there will be no changes as dramatic as those experienced since the 1930s. Trees will grow somewhat taller, and many of the locusts, pines, sassafras, and other pioneers will give way to the more shade-tolerant oaks, hickories, and maples. Ridgetop forests will remain stunted as the elements buffet them, but trees in coves will continue toward their giant potential. Some forest creatures

> **Do You Know:**
> What effect global warming may have on the flora and fauna of the park?
> What programs are offered at the Great Smoky Mountains Institute at Tremont?
> The significance of the park's Natural History Collection?

will probably increase. Woodpeckers, for instance, will benefit as the aging forest produces more insect-infested wood. But animal populations in general will probably decline somewhat as the closing canopy shades out vegetation near the forest floor.

Global Warming

Except for nuclear war or a collision with an asteroid, no force has more potential to damage our planet's web of life than global warming. A decade ago, the idea that the planet was warming up as a result of human activity was largely theoretical. Not anymore. As authoritative reports issued in 2005 and 2007 by the United Nations–sponsored Intergovernmental Panel on Climate Change makes plain, the trend toward a warmer world has unquestionably begun and carbon dioxide produced by humans is largely to blame. The Earth has been warming at a rate of 0.36 degrees Fahrenheit per decade for the past 30 years, and the 1990s were the hottest decade on record. The year 2006 was the warmest year in a century. The overall temperature is the warmest in the current interglacial period, which began about 12,000 years ago. After analyzing data going back at least two decades on everything from air and ocean temperatures to the spread and retreat of wildlife, the IPCC asserts that the slow but steady warming has had an impact on no fewer than 420 physical processes and animal and plant species on all continents.

Increased levels of carbon dioxide foster the rapid growth of many types of vines such as honeysuckle and poison ivy. Poison ivy vines grew more than twice as much per year under elevated carbon dioxide levels as they did in unaltered air. Vines don't spend much of their carbon harvest on trunks or other supports, so the carbon windfall can go directly into new leaves, which collect yet more carbon and sunlight. An increased abundance of vines can potentially choke out trees and change forest dynamics.

According to the National Climatic Data Center in Asheville, North Carolina, 2006 was the warmest in the United States since record keeping began in 1895. The average temperature for the 48 contiguous United States was 50.6 degrees F, or 2.2 degrees above the average of the 20th century.

Atmospheric molecules of carbon dioxide, water, nitrous oxide, methane, and chlorofluorocarbons are among the main players in interactions that affect global temperature. Collectively, these gases act like the panes of glass in a greenhouse—hence the name, "greenhouse gases." Wavelengths of visible light pass through these gases to Earth's surface, which absorbs them and emits longer, infrared wavelengths—heat. Greenhouse gases impede the escape of heat energy from Earth into space. How? The gaseous molecules absorb the longer wavelengths, then radiate much of it back toward Earth.

Growing seasons are getting longer, about six days longer in the northeastern United States, about 25 days longer in the Northwest. Glaciers are disappearing from mountaintops around the globe. Africa's two highest mountains, Mount Kilimanjaro and Mount Kenya, will lose their ice cover within 25 to 50 years if deforestation and industrial pollution are not stopped. Sea levels are rising. Coral reefs are dying off as the seas get too warm for comfort. Drought is the norm in parts of Asia and Africa. The Arctic permafrost is starting to melt. Lakes and rivers in colder climates are freezing later and thawing earlier each year. Plants and animals are shifting their ranges poleward and to higher altitudes, and migration patterns for animals as diverse as polar bears, butterflies, and beluga whales are being disrupted. A study of 1,700 plant, insect, and animal species found poleward migration of about four miles (6 km) per decade and vertical migration in alpine regions of about 20 feet (6 m) per decade in the second half of the 20th century. The range of the Edith's Checkerspot butterfly of western North America has moved almost 60 miles north in 100 years. When birds' arrival and insects' appearance get out of sync, it can mean scarce food for hatchlings. Delays in spring arrival by migratory birds may lead to increased competition for nest sites with species arriving earlier.

Faced with these hard facts, scientists no longer doubt that global warming is happening. Although some researchers dispute specific aspects of global warming, 2,500 of the world's leading climate scientists from 113 countries who serve on the IPCC have concluded that global warming is real and that carbon dioxide is largely to blame. The obvious conclusion: temperatures will keep going up. Unfortunately, they may be rising faster and heading higher than anyone expected. The 2007 report stated that warming during the last 100 years was 0.74 degrees C (1.3° F) with most of the warming occurring during the past 50 years. The warming for the next 20 years is projected to be 0.2 degrees C (0.3° F) per decade. Probable temperature rise by the end of the century will be between 1.8 degrees C and 4 degrees C (3.2–7.2° F) That may not seem like much, but consider that it took only a 9 degree F shift to end the last ice age. Entire climatic zones might shift dramatically, making central Canada look more like central Illinois, Georgia more like Guatemala.

Scientists have been warning for decades that carbon dioxide and other "greenhouse" gases from power plants and vehicle exhaust are warming the planet and raising the seas. They say the best way to minimize the damage is to drastically reduce smokestack and tailpipe emissions. The 2007 IPCC report states that continued warming is likely if the concentration of carbon dioxide in the atmosphere doubles from 280 parts per million, which was the average for many centuries preceding the Industrial Revolution. The carbon dioxide concentration is now roughly 380 parts per million, and many climate experts say it will be extremely difficult to avoid hitting levels of 450 or 550 parts per million, or higher, later this century, given growth in populations and fuel use and the lack of nonpolluting alternatives that can be exploited at a sufficient scale to replace fossil fuels.

Scientists reported in 2006 that the impact of melting from Antarctica's ice sheets has been underestimated. The 2007 IPCC report states that Arctic summer sea ice is likely to disappear in the second half of the current century. The melting will exacerbate the effects of global warming and play a major role in submerging many coastal communities if nothing is done to curb the emissions. No one is sure of the extent of the melting or the timing of its effects. But the researchers say that with the warming climate, melting ice sheets in Greenland, the Arctic, and Antarctica could inundate coastal areas around the world. Earth's average temperature has increased about 1 degree F over the past 30 years, a relatively rapid rise, and sea levels are rising about an inch a decade. Predictions are that sea levels could rise up to three feet over the next century.

A potential problem for many species is the possibility of fairly rapid (50–100 years) changes in climate. Wildlife in even the best-protected and best-managed reserves could be depleted in a few decades if such changes in climate take place.

Global warming will also change the makeup and location of many of the world's forests. Forests in temperate and subarctic regions would move toward the poles or to higher altitudes, leaving more grassland and shrubland in their wake. Tree species, however, move slowly through the growth of new trees along forest edges—typically about 0.9 kilometer (0.5 mile) per year or nine kilometers (five miles) per decade. According to the 1995 report of the IPCC, mid-latitude climate zones are projected to shift northward by 550 kilometers (340 miles) over the next century. At that rate some tree species such as beech might not be able to migrate fast enough and would die out. According to the IPCC, over the next century "entire forest types might disappear, including half of the world's dry tropical forests."

Climate change would lead to reductions in biodiversity in many areas. Large-scale forest diebacks may sharply reduce populations of some species, especially those with specialized niches, and cause mass extinction of plant and animal species that couldn't migrate to new areas. Extensive areas of dead timber could provide fuel for an increased number of forest fires. Fish would die as temperatures soared in streams and lakes and as lowered water levels concentrated pesticides. Any shifts in regional climate would threaten many parks, wildlife reserves, and wilderness areas, wiping out many current efforts to stem the loss of biodiversity. Worst of all, this increase in temperatures is happening at a pace that outstrips anything the Earth has seen in the past 100 million years.

Species are not equally susceptible to becoming extinct. Species with limited habitats become extinct easily simply because they have so little available habitat to destroy or disturb.

Researchers have documented shifts in the ranges of many butterflies. One study looked at 35 species of nonmigratory butterflies whose ranges extended from northern Africa to northern Europe. The scientists found that two-thirds of the species had shifted their home ranges northward by 20 to 150 miles during the last century. Climate change, either directly or indirectly, may be the culprit in most cases of declining amphibian populations. Climate change makes frogs more vulnerable to the fungus *Batrachochytrium dendrobatidis*. The home habitat of the golden toad in Costa Rica moved up the mountain until "home" disappeared entirely. What troubles scientists especially is that if we are only in the early stages of warming, all these lost and endangered animals might be just the first of many to go. One study estimates that more than a million species worldwide could be driven to extinction by the year 2050.

What effect might global warming have on the Smokies? It could mean the elimination of the spruce-fir forest within the park. If this forest type should disappear, it would affect a great many other plant and animal species, many of which reach their southernmost distributions in or near the Smokies—the northern flying squirrel, the northern water shrew, the spruce-fir moss spider, and others. Salamander distribution would be affected. Warmer stream temperatures could diminish native trout populations. Bird distribution would be affected. Populations of birds that engage in vertical migrations, such as the Carolina chickadee and the dark-eyed junco, would need to travel much greater distances to find suitable nesting and wintering habitat. If the lower elevations became warmer, it would favor such species as the anole and other terrestrial reptiles which could expand their ranges. It would also alter bird migration patterns, plant distributions, flowering cycles, and breeding patterns. Climate change may also promote the spread of invasive pests such as the hemlock wooly adelgid. Art Stupka was interested in year-to-year changes and recorded shifts in blooming dates for the same trees, some over a period of 17 years. Unfortunately, the work was not continued after his retirement. Now such work is very important, but at that time, global climate change was not an issue. In 1999, counts of buds, blooms, and spent blossoms on tagged rosebay rhododendron at several locations in the park were begun. Counts are made at least two times per year. Results are then averaged and full bloom season documented with photographs. This species was chosen because it is found over hundreds of square kilometers in the Smokies and because the blooming of these shrubs is cyclic. What controls and synchronizes these millions of shrubs is unknown. Other plants like ramps also seem to regularly have greater sexual reproduction in odd-numbered years. Apparently climatic factors in combination with internal food storage exhaustion play a major part in the synchrony. The study of such periodic biologic phenomena is known as *phenology*.

While there will always be trees, wildflowers, birds, and other wildlife in the park, the species that are present as well as the flowering and breeding seasons may be significantly different for future generations than they are now.

Though the views from Chimneys or Clingmans Dome do not tell us, nature creates and nature destroys. We, too, are involved, for we are as inevitably tied to our environment, worldwide though it may be, as any salamander on a log or any lichen on a tree. If our environment becomes unfit for us, we will, as surely as the red-cheeked salamander or the spruce-fir moss spider in their shrinking habitat, face extinction.

Within the park, research and education will continue at two existing facilities (Great Smoky Mountains Institute at Tremont and the Appalachian Highlands Science Learning Center at Purchase Knob) and at one new facility (Twin Creeks Science Center). The park's natural history collection will continue to expand and will be of increasing value to a wide range of researchers.

Great Smoky Mountains Institute at Tremont

www.gsmit.org
(865) 448-6709

The mission of the Great Smoky Mountains Institute at Tremont is to connect people and nature through in-depth programs designed to nurture appreciation of Great Smoky Mountains National Park, celebrate diversity, and foster stewardship. The institute hosts over 5,000 students (children and adults) each year in school programs, summer camps, and adult workshops.

The institute began in 1969 as the Tremont Environmental Center and was managed by Maryville College. It was originally a Job Corps facility constructed in the mid-1960s. The current office, dormitory, and activity center are all original to that facility, although their uses are obviously different. Maryville College terminated its contract around 1980, and the facility sat unused for several years. In 1983 the Great Smoky Mountains Natural History Association (now known as the Great Smoky Mountains Association) took over management of the facility, renamed it Great Smoky Mountains Institute at Tremont (GSMIT), and hired Ken Voorhis as director. In 2000 the institute parted amicably from the association in order to become an independent nonprofit organization and allow growth in new directions. The institute continues to maintain a close relationship with the association as well as the National Park Service. The institute operates under an agreement with the NPS whose members sit on the board of directors and have input into all operational and programmatic activities.

Appalachian Highlands Science Learning Center at Purchase Knob

www.nps.gov/grsm/pksite/index.htm
www.nps.gov/grsm/pksite/history.htm
http://www.nps.gov/grsm/pksite/pk-birds-annotated-06.pdf—bird list
http://www.nps.gov/grsm/pksite/images/butterflies.pdf—butterfly list
(828) 926-6251

The Appalachian Highlands Science Learning Center is based on 535 acres in Haywood County, North Carolina, contiguous with the rest of Great Smoky Mountains

National Park. It includes Purchase Knob (5,086 ft. elevation), a historic cabin, and two buildings that have been converted into offices, laboratory space, conference rooms, classroom space, and housing for visiting scientists and their students. The buildings and land were donated in 2000 by Kathryn McNeil and Voit Gilmore who had owned the property since 1964 and had built a summer home on it. The cabin was built by John and Emily Ferguson in 1884–85. They purchased what was then 447 acres for $447 and a horse. They farmed the land, which became a cattle pasture in the mid-1900s. The McNeil/Gilmores purchased the property in 1964, built the two modern buildings, and planted Fraser firs for a Christmas tree plantation. Much of the forest on the Purchase Knob site was never cut over, although there was some selective cutting, and the chestnut blight hit the area hard. The park is maintaining most of the existing field as field, managing five areas as scrub habitat, and otherwise letting the forest mature. Within the next five years, the park plans to conduct a public planning process to design a plan for the long-term management of the property, but it will be maintained as a field station for researchers while they are working in the park or the southern part of the Blue Ridge Parkway, as well as for educational programs based on the research. In 2001 Purchase Knob became the home of one of five initial Research Learning Centers created by Congress to support research in the national parks and to transmit the information generated to the public. There are now 18 centers throughout the nation's national parks.

Twin Creeks Science Center

Construction began in early 2006 for this much-needed building. It is a one-story building with a large open area that can be reconfigured as needs change. The building is very energy-efficient and includes a small water-powered electrical generator that will help supply the facility's power needs. An environmental education room adjoins the science area. There are a small number of private offices for permanent employees, a large curatorial space, labs, an insect rearing room, GIS and computer facilities, and storage. There is considerable covered porch space, a large carport to protect government vehicles and to allow loading of vehicles the night before fieldwork, and a screen/glass porch. A huge topographic map of the park is located along one wall.

Natural History Collection

The park's natural history collection, which had been maintained in climate-controlled facilities in the basement of the Sugarlands Visitor Center, was moved into the Twin Creeks Science Center in 2007. The collection, which was begun by Arthur Stupka, currently (June 2006) consists of 40,926 catalog records (11,313 plants, 24,664 insects, 208 arachnids, 1,767 amphibians, 441 reptiles, 164 birds, and 830 mammal specimens in addition to geology and paleontology specimens). The current curator, Adriean Mayor, stated that it is the largest natural history collection in the National Park Service. The insect collection is the largest in Tennessee; the plant collection is probably the third-largest in Tennessee. The collection has served and will continue to serve as a valuable resource to researchers worldwide.

Two other educational programs for children and families are Parks as Classrooms and the Smoky Mountain Field School.

Parks as Classrooms

www.nps.gov/grsm/gsmsite/parksasclass.html
(865) 436-1292

This is a day-use education program that provides outdoor learning experiences in the park for students in grades K–8. The program offers curriculum-based, interdisciplinary lessons that weave together Great Smokies themes with Tennessee and North Carolina curricula. It features park rangers as the subject experts and primary instructors with assistance from the classroom teachers. In addition to the on-site learning experience, the units include a pre-visit materials package and post-site lesson plans for use in the classroom.

Smoky Mountain Field School

www.outreach.utk.edu/smoky
(865) 974-0150

An educational outreach program of the University of Tennessee designed for adults and families, the Field School offers hiking and other outdoor programs which run from four hours to two days. Programs are frequently held on the weekends and cover various aspects of natural and cultural history, including wildflowers, fireflies, black bears, Cherokee history, and orienteering. The Field School strives to provide knowledgeable instructors who are recognized experts in their fields.

The following organizations provide funding for a wide variety of park activities including research, educational programs, cultural and natural history activities, publications, and maintenance and improvements to visitor facilities, campsites, and trails.

Great Smoky Mountains Conservation Association

This association was organized in December 1923 and must be considered the most important of all organizations that took an active part in the creation of the park. Its original stated purpose was "to establish a National Park in the Great

Smoky Mountains and to protect and promote its interest before and after its completion." Among its founding members were Mr. and Mrs. Willis Davis and David Chapman, three of the more vocal Knoxville supporters of a park. The association continues to provide graduate-level scholarships for park research.

The Great Smoky Mountains Association

www.SmokiesInformation.org

This Association was organized in 1953 with a Tennessee state charter and official designation as a cooperative society by the Department of the Interior. The association operates as an educational nonprofit corporation for the benefit of the interpretive program in the park. Formerly known as the Great Smoky Mountains Natural History Association, the name was changed to the Great Smoky Mountains Association in 2003. Since 1953, the association has been supporting the educational, scientific, and historical efforts of the National Park Service through cash donations and in-kind services amounting to over $16 million. In 2006 alone, the association provided $1,644,215 worth of assistance. Projects supported include resource management projects such as the elk reintroduction and Gatlinburg bear warden; funding long-term scientific research projects such as the All Taxa Biodiversity Inventory (ATBI); sponsoring the Spring Wildflower Pilgrimage; sponsoring living history demonstrations and environmental education programs; funding seasonal rangers who conduct walks and talks in the park; and publishing the park newspaper, *Smokies Guide*, as well as books and pamphlets on the Smokies.

Friends of Great Smoky Mountains National Park

www.friendsofthesmokies.org

Friends of the Smokies is a nonprofit organization that assists the National Park Service by raising funds and public awareness and providing volunteers for needed projects. Since 1993, Friends has given over $12 million for park projects and programs. For 2006 alone, the total amount of funding was $1,092,162. These donations have helped:

> Protect elk, bear, brook trout, and other wildlife.
> Assist in efforts to control the hemlock woolly adelgid.
> Provide continuing support for the ATBI.
> Provide support for the Twin Creeks Science and Education Center.
> Improve trails, campsites, and backcountry shelters.
> Support educational programs for schoolchildren.
> Improve visitor facilities.
> Fund special educational services like the official park movie.
> Preserve log cabins and other historic structures.

Our national parks are gifts from prior generations that Americans cherish. They are treasures to be treated with utmost care so that future generations, as well as our own, may experience them as our forebears did. Regardless of which national park you may visit, the ability of human beings to reflect in the natural quiet of those places is fundamental to the experience.

APPENDIX I

WHERE IS THAT?

PARK LOCALITIES REFERENCED IN TEXT

ABRAMS CREEK 857–3,075 ft. (261–938 m)
    Near western boundary of park (Tennessee)
ALBRIGHT GROVE 3,100–3,350 ft. (946–1,022 m)
    Area of virgin forest on Maddron Bald Trail (Tennessee)
ALUM CAVE BLUFFS 1,200 ft. (366 m)
    South slope of Mount LeConte (Tennessee)
ALUM CAVE TRAIL 3,800–6,300 ft. (1,159–1,922 m)
    South slope of Mount LeConte (Tennessee)
ANDREWS BALD 5,800 ft. (1,769 m)
    South of Clingmans Dome (North Carolina)
APPALACHIAN TRAIL mostly 5,000–6,000 ft. (1,525–1,830 m)
    Approximately along Tennessee–North Carolina state line
ASH CAMP BRANCH 3,700–4,400 ft. (1,150–1,540 m)
    Tributary of Fish Camp Prong (Tennessee)
BASKINS CREEK FALLS 2,130 ft. (650 m)
    Along Baskins Creek between Roaring Fork Motor Nature Trail and Cherokee Orchard Road (Tennessee)
BALSAM MOUNTAIN 5,410 ft. (1,650 m)
    In southeastern portion of park near the Qualla Boundary (North Carolina)
BAXTER CREEK 1,800–4,600 ft. (560–1,400 m)
    Tributary of Big Creek (North Carolina)
BEAR CREEK 1,800–4,100 ft. (550–1,250 m)
    Tributary of Forney Creek (North Carolina)
BIG CREEK approximately 1,500–4,900 ft. (458–1,495 m)
    Near northeastern boundary of park (North Carolina)
BIG FORK RIDGE 3,300–3,800 ft. (1,000–1,150 m)
    Ridge off Balsam Mountain between Cataloochee Balsam and Polls Gap near eastern boundary of park (North Carolina)
BLANKET MOUNTAIN 4,609 ft. (summit) (1,406 m)
    Southwest of Elkmont (Tennessee)

BLOWHOLE CAVE approximately 1,750 ft. (534 m)
   In Whiteoak Sink (Tennessee)
BOOGERMAN LOOP TRAIL 2,900–3,300 ft. (900–1,000 m)
   On eastern side of Big Fork Ridge in Cataloochee (North Carolina)
BOULEVARD PRONG 2,600–5,200 ft. (800–1,575 m)
   Tributary of Porters Creek (Tennessee)
BREAKNECK RIDGE 4,900–5,400 ft. (1,500–1,650 m)
   Spur extending west from Hyatt Ridge (North Carolina)
BRYSON CITY 1,700 ft. (519 m)
   On south-central boundary of park (North Carolina)
BUCK FORK 2,600–5,400 ft. (800–1,650 m)
   Tributary of Middle Prong Little Pigeon River (Tennessee)
BULL CAVE 1,900 ft. (580 m)
   Northern boundary of park, near Rich Mountain (Tennessee)
BULLHEAD BRANCH approximately 1,500–1,800 ft. (460–550 m)
   Tributary of West Prong Little Pigeon River (Tennessee)
CADES COVE mostly 1,800–1,900 ft. (549–580 m)
   Western part of park (Tennessee)
CANNON CREEK 2,460–5,080 ft. (750–1,550 m)
   Flows northeast from Mount Le Conte to Porters Creek (Tennessee)
CATALOOCHEE approximately 2,600 ft. (793 m)
   Cove near eastern boundary of park (North Carolina)
CHAMBERS CREEK 1,800–4,300 ft. (550–1,300 m)
   Tributary of Fontana Lake southeast of Welch Ridge (North Carolina)
CHARLIES BUNION 5,375 ft. (1,639 m)
   On Appalachian Trail, northeast of Newfound Gap (Tennessee–North Carolina)
CHEROKEE approximately 1,900 ft. (580 m)
   Town on Qualla Boundary adjacent to southern boundary of park (North Carolina)
CHEROKEE ORCHARD 2,600 ft. (793 m)
   Four miles southeast of Gatlinburg (Tennessee)
CHILHOWEE MOUNTAIN approximately 1,500–2,700 ft. (458–824 m)
   Outside park, just beyond northwestern boundary (Tennessee)
CHIMNEY TOPS 4,755 ft. (1,450 m)
   Along transmountain road (Tennessee)
CHIMNEYS PICNIC AREA 2,700 ft. (824 m)
   Along Newfound Gap Road, six miles south of Gatlinburg (Tennessee)
CLINGMANS DOME 6,643 ft. (summit) (2,026 m)
   Along state line at head of Forney Ridge, highest point in park (Tennessee–North Carolina)

COSBY approximately 1,400 ft. (427 m)
: Town north of park boundary near northeast corner of park (Tennessee)

COSBY CAMPGROUND approximately 2,400 ft. (732 m)
: Northeast corner of park (Tennessee)

COSBY CREEK approximately 1,650–4,200 ft. (503–1,281 m)
: Near northeastern boundary of park (Tennessee)

COVE MOUNTAIN 4,091 ft. (1,248 m)
: Northern boundary of park, west of Gatlinburg, Tennessee

DALTON GAP 3,100 ft. (950 m)
: Along state line ridge near southwestern boundary of park (North Carolina)

DAVENPORT GAP 1,902 ft. (580 m)
: Northeast boundary of park at junction with state line (Tennessee–North Carolina)

DEEP CREEK approximately 1,792–4,000 ft. (547–1,220 m)
: North of Bryson City (North Carolina)

DOUBLE GAP BRANCH 3,100–4,260 ft. (950–1,300 m)
: Tributary of Caldwell Fork (North Carolina)

DOUBLE SPRINGS GAP 5,590 ft. (1,704 m)
: On state-line ridge east of Silers Bald (Tennessee–North Carolina)

DUNN CREEK 1,965–5,085 ft. (600–1,550 m)
: North of Mount Guyot (Tennessee)

EAGLE CREEK approximately 1,700–2,500 ft. (519–763 m)
: North of Fontana Lake (North Carolina)

ELKMONT 2,146 ft. (655 m)
: On Little River, southwest of park headquarters (Tennessee)

ENLOE CREEK 3,600–5,200 ft. (1,100–1,650 m)
: Tributary of Raven Fork (North Carolina)

FIGHTING CREEK GAP 2,320 ft. (708 m)
: On Little River Road, four miles west of park headquarters (Tennessee)

FONTANA LAKE between 1,700 and 1,800 ft. (519–549 m)
: Along southwestern boundary of park (North Carolina)

FONTANA VILLAGE approximately 2,000 ft. (610 m)
: Near southwestern boundary of park (North Carolina)

FOOTHILLS PARKWAY
: Unfinished two-lane highway that circles the Tennessee side of the park

FORGE CREEK approximately 1,800–2,000 ft. (549–610 m)
: Flows south from the vicinity of Gregory Bald to Cades Cove (Tennessee)

FORNEY CREEK approximately 1,600–5,700 ft. (488–1,739 m)
: Flows southwest from vicinity of Clingmans Dome to Fontana Reservoir (North Carolina)

GABES CREEK 2,450–4,100 ft. (750–1,250 m)
   Tributary of Greenbrier Creek (Tennessee)
GATLINBURG 1,293 ft. (394 m)
   Town on north-central boundary of park (Tennessee)
GREENBRIER 1,680 ft. (512 m)
   Approximately 10 miles east of Gatlinburg (Tennessee)
GREENBRIER PINNACLE 4,805 ft. (summit) (1,466 m)
   East of Greenbrier, near northeastern boundary of park (Tennessee)
GREGORY BALD 4,948 ft. (1,509 m)
   On state line ridge, southwest of Cades Cove (Tennessee–North Carolina)
GREGORY CAVE 1,900 ft. (575 m)
   Off Cades Cove Loop Road (Tennessee)
HALF ACRE RIDGE approximately 4,100 feet (1,250 m)
   Short spur extending northwest from Cataloochee Divide (North Carolina)
HAPPY VALLEY 1,332 ft. (406 m)
   On western boundary of park near Chilhowee Mountain (Tennessee)
HAZEL CREEK approximately 2,100–5,150 ft. (641–1,571 m)
   Flows into Fontana Reservoir between Eagle Creek and Forney Creek (North Carolina)
HUGGINS CREEK 2,800–5,250 ft. (850–1,600 m)
   Tributary of Forney Creek (North Carolina)
HUGHES RIDGE mostly 4,000–5,500 ft. (1,220–1,677 m)
   From Pecks Corner, on state line, south into the Qualla Boundary (North Carolina)
INDIAN CAMP CREEK approximately 1,850–4,850 ft. (564–1,479 m)
   North of Old Black Mountain in northeastern section of park (Tennessee)
INDIAN GAP 5,266 ft. (1,606 m)
   West of Newfound Gap, along road to Clingmans Dome (Tennessee–North Carolina)
JONES BRANCH 3,100–5,075 ft. (950–1,550 m)
   Tributary of Maddron Creek (Tennessee)
LAUREL CREEK approximately 1,200–1,800 ft. (366–549 m)
   Tributary of the Middle Prong of Little River; along spur road to Cades Cove (Tennessee)
LE CONTE CREEK 1,300 ft. (at mouth) (397 m)
   Flows from Mount LeConte into Gatlinburg, Tennessee.
LE CONTE LODGE approximately 6,300 ft. (1,922 m)
   Near summit of Mount LeConte, at junction of trails (Tennessee)
LITTLE GREENBRIER 1,650–2,450 ft. (500–750 m)
   Near Metcalf Bottoms (Tennessee)

LITTLE RIVER approximately 1,100–5,350 ft. (336–708 m)
    Originates north of Mount Buckley; leaves park at northwestern boundary near Townsend (Tennessee)

LOOK ROCK 2,650 ft. (808 m)
    Near western boundary on Chilhowee Mountain (Tennessee)

LOW GAP 4,242 ft. (1,294 m)
    On state line, north of Cosby Knob, in northeastern part of park (Tennessee)

LOW GAP TRAIL 2,500–4,242 ft. (763–1,294 m)
    Trail from Cosby Campground to Low Gap (Tennessee)

LUFTEE KNOB 6,234 ft.
    Along Balsam Mountain Trail near state line (North Carolina)

LYNN CAMP PRONG 2,450–4,200 ft. (750–1,300 m)
    Tributary of Middle Prong Little River (Tennessee)

MADDRON BALD TRAIL 3,100–5,500 ft. (946–1,678 m)
    Between Indian Camp Creek and Snake Den Mountain in Cosby section of park (Tennessee)

MALONEY POINT 2,300 ft. (700 m)
    Overlook on Little River Road, 0.3 mile east of Fighting Creek Gap (Tennessee)

MANNIS BRANCH 2,000–3,000 ft. (600–900 m)
    Tributary of Little River on north side of Meigs Mountain (Tennessee)

MESSER FORK 2,950–4,400 ft. (900–1,350 m)
    Tributary of Rough Fork (North Carolina)

METCALF BOTTOMS 1,679 ft. (512 m)
    Picnic area along Little River, two miles above the Sinks bridge (Tennessee)

MOUNT BUCKLEY 6,582 ft. (summit) (2,008 m)
    One mile west of Clingmans Dome on state line (Tennessee–North Carolina)

MOUNT CAMMERER 5,025 ft. (summit) (1,533 m)
    Near state line in extreme northeast corner of park (Tennessee)

MOUNT COLLINS 6,188 ft. (summit) (1,888 m)
    On state line between Indian Gap and Clingmans Dome (Tennessee–North Carolina)

MOUNT GUYOT 6,621 ft. (summit) (2,019 m)
    On state line east of Greenbrier (Tennessee–North Carolina)

MOUNT KEPHART approximately 6,200 ft. (summit) (1,891 m)
    On state line three miles northeast of Newfound Gap (Tennessee–North Carolina)

MOUNT LE CONTE 6,593 ft. (summit) (2,011 m)
    Third-highest peak in park; southeast of Gatlinburg (Tennessee)

MOUNT LOVE 6,420 ft.
: Fifth-highest peak in park; near Clingmans Dome (Tennessee–North Carolina)

MOUNT MINGUS 5,800 ft. (1,769 m)
: North of Indian Gap (Tennessee)

MOUNT STERLING 5,835 ft. (summit) (1,780 m)
: Near eastern boundary of park; south of town of Mount Sterling (North Carolina)

MOUSE CREEK 2,300–4,900 ft. (700–1,500 m)
: Tributary of Big Creek (North Carolina)

NEWFOUND GAP 5,040 ft. (1,537 m)
: Highest point on Newfound Gap Road; on state line and along Appalachian Trail (Tennessee–North Carolina)

NEWFOUND GAP ROAD 1,300–5,040 ft. (397–1,537 m)
: Road from Gatlinburg, Tennessee, to Cherokee, North Carolina, via Newfound Gap (Tennessee–North Carolina)

NOLAND CREEK approximately 2,600–5,200 ft. (793–1,586 m)
: Between Forney Ridge and Noland Divide (North Carolina)

OCONALUFTEE RIVER approximately 1,950–3,050 ft. (595–930 m)
: Flows along Newfound Gap Road to southern boundary of park at Cherokee (North Carolina)

OLD BLACK MOUNTAIN 6,356 ft. (summit) (1,939 m)
: On state line, one mile north of Mount Guyot (Tennessee–North Carolina)

PARK HEADQUARTERS BUILDING 1,460 ft. (445 m)
: Administration building, two miles south of Gatlinburg (Tennessee)

PILKEY CREEK 1,800–4,100 ft. (550–1,250 m)
: Tributary of Fontana Lake south of Welch Ridge (North Carolina)

PORTERS CREEK 1,680 ft. (mouth) (512 m)
: Tributary of Middle Prong, Little Pigeon River, south of Greenbrier (Tennessee)

PROCTOR 1,700 ft. (519 m)
: Abandoned town north of Fontana Lake along Hazel Creek (North Carolina)

PURCHASE KNOB 5,075 ft. (summit) (1,550 m)
: Near Cataloochee; site of Appalachian Highlands Science Learning Center (North Carolina)

RABBIT CREEK 1,300–2,600 ft. (397–793 m)
: Tributary of Abrams Creek, west of Cades Cove (Tennessee)

RAINBOW CAVE approximately 1,750 ft. (534 m)
: In Whiteoak Sink (Tennessee)

RAINBOW FALLS TRAIL 2,581–6,300 ft. (787–1,922 m)
    From Cherokee Orchard to Mount LeConte via Rocky Spur (Tennessee)
RAVENSFORD 2,100 ft. (640 m)
    Near junction of Raven Fork and Oconaluftee River; in vicinity of Oconaluftee Visitor Center (North Carolina)
ROARING FORK approximately 1,300 ft. (mouth) (397 m)
    Enters West Prong of Little Pigeon River in Gatlinburg; originates north of Cliff Top on Mount LeConte (Tennessee)
RUSSELL FIELD 4,230 ft. (1,290 m)
    On state line southeast of Cades Cove
SAMS CREEK 2,100–4,300 ft. (650–1,300 m)
    Branch of Thunderhead Prong
SCHOOLHOUSE GAP approximately 2,000 ft. (610 m)
    On northern boundary of park on old road between Whiteoak Sink (vicinity) and Dry Valley in Tuckaleechee Cove (Tennessee)
SCOTT GAP CAVE approximately 1,800 ft. (523 m)
    In Whiteoak Sink
SILERS BALD 5,620 ft. (1,714 m)
    On state line west of Clingmans Dome (Tennessee–North Carolina)
SKUNK RIDGE 2,296 ft. (700 m)
    Near western boundary of park (Tennessee)
SMOKEMONT 2,198 ft. (670 m)
    On Newfound Gap Road above Oconaluftee Visitor Center (North Carolina)
SPENCE FIELD approximately 5,000 ft. (1,525 m)
    Just west of Thunderhead on state line (Tennessee–North Carolina)
STEELTRAP CREEK 3,900–5,400 ft. (1,200–1,650 m)
    Tributary of Forney Creek (North Carolina)
STYX BRANCH 4,100–6,000 ft. (1,250–1,800 m)
    Tributary of Alum Cave Creek (Tennessee)
SUGARLANDS 1,500–2,700 ft. (458–824 m)
    Valley from near Sugarlands Visitor Center to Chimneys picnic area (Tennessee)
TABCAT CREEK 900–1,600 ft. (275–500 m)
    Tributary of Chilhowee Lake (Tennessee)
THUNDERHEAD 5,530 ft. (summit) (1,687 m)
    On state line southeast of Cades Cove (Tennessee–North Carolina)
TOWNSEND 1,100 ft. (336 m)
    Town in Tuckaleechee Cove, two miles north of park boundary (Tennessee)
TREMONT 1,925 ft. (507 m)
    On Middle Prong of Little River near junction with Lynn Camp Prong (Tennessee)

TWENTYMILE CREEK 1,313–4,150 ft. (400–1,266 m)
: Flows into Cheoah Lake on park boundary west of Fontana Dam (North Carolina)

TWIN CREEKS 2,000 ft. (600 m)
: Off Cherokee Orchard Road along LeConte Creek. Former residence of park superintendent and Twin Creeks Natural Resource Center. Site of Twin Creeks Science Center.

WALKER CAMP PRONG 3,800–5,600 ft. (1,150–1,700 m)
: Tributary of West Prong Little Pigeon River (Tennessee)

WEBB CREEK 1,640–4,400 ft. (500–1,350 m)
: Tributary of Middle Prong Little Pigeon River (Tennessee)

WELCH BRANCH approximately 1,800 ft. (mouth) (550 m)
: Tributary of Fontana Lake near Chambers Creek (North Carolina)

WEST PRONG OF LITTLE PIGEON RIVER approximately 1,300–4,600 ft. (397–1,403 m)
: Main stream along Newfound Gap Road from Gatlinburg, Tennessee, to near Newfound Gap (Tennessee)

WHITEOAK SINK approximately 1,750 ft. (534 m)
: Small cove just inside park boundary, northeast of Cades Cove (Tennessee)

WINDING STAIRS BRANCH 2,450–3,600 ft. (750–1,100 m)
: Tributary of Cataloochee Creek (North Carolina)

APPENDIX II

CHECKLIST OF TREES, SHRUBS, AND VINES REFERENCED IN TEXT

A date next to an entry indicates the year the species was discovered as part of recent ATBI research.

Allegheny serviceberry (*Amelanchier laevis*)
Alternate-leaf dogwood (*Cornus alternifolia*)
American beech (*Fagus grandifolia*)
American chestnut (*Castanea dentata*)
American elm (*Ulmus americana*)
American holly (*Ilex opaca*)
American mistletoe (*Pharadendron leucarpum*)
American mountain-ash (*Sorbus americana*)
American sycamore (*Platanus occidentalis*)
Apple (*Malus pumila*)
Bitternut hickory (*Carya cordiformis*)
Blackberry (*Rubus* sp.)
Black cherry (*Prunus serotina*)
Black gum (*Nyssa sylvatica*)
Black locust (*Robinia pseudoacacia*)
Black oak (*Quercus velutina*)
Blueberry (*Vaccinium* sp.)
Bush honeysuckle (*Diervilla* sp.)
Butternut (*Juglans cinerea*)
Carolina hemlock (*Tsuga caroliniana*)
Carolina rhododendron (*Rhododendron caroliniana*)
Catawba rhododendron (*Rhododendron catawbiense*)
Chestnut oak (*Quercus prinus*)
Cinnamon clethra or Mountain pepper-bush (*Clethra acuminata*)
Common alder (*Alnus serrulata*)
Climbing euonymus (*Euonymus fortunei*) [exotic]
Cucumber tree (*Magnolia acuminata*)
Devil's walkingstick (*Aralia spinosa*)

Dog-hobble (*Leucothoe fontanesiana*)
Dutchman's pipe (*Aristolochia macrophyllia*)
Eastern hemlock (*Tsuga canadensis*)
Eastern leatherwood (*Dirca palustris*)
English ivy (*Hedera helix*) [exotic]
Flame azalea (*Rhododendron calendulaceum*)
Flowering dogwood (*Cornus florida*)
Fraser fir (*Abies fraseri*)
Fraser magnolia (*Magnolia fraseri*)
Grape (*Vitis* sp.)
Green ash (*Fraxinus pennsylvanica*)
Hawthorn (*Crataegus* sp.)
Heart's-a-bustin' (*Euonymus americanus*)
Hobblebush or Witchhobble (*Viburnum lantanoides*)
Hop-hornbeam (*Ostrya virginiana*)
Huckleberry (*Gaylussacia* sp.)
Ironwood or American hornbeam (*Carpinus caroliniana*)
Japanese honeysuckle (*Lonicera japonica*) [exotic]
Japanese spiraea (*Spiraea japonica*) [exotic]
Kudzu (*Pueraria montana* var. *lobata*) [exotic]
Mimosa (*Albizia julibrissin*) [exotic]
Mountain laurel (*Kalmia latifolia*)
Mountain maple (*Acer spicatum*)
Mountain stewartia (*Stewartia ovata*)
Multiflora rose (*Rosa multiflora*) [exotic]
Northern red oak (*Quercus rubra*)
Norway spruce (*Picea abies*) [exotic]
Persimmon (*Diospyros virginiana*)
Pin cherry (*Prunus pensylvanica*)
Pignut hickory (*Carya glabra*)
Pitch pine (*Pinus rigida*)
Privet (*Ligustrum vulgare*) [exotic]
Red elderberry (*Sambucus pubens*)
Red maple (*Acer rubrum*)
Red spruce (*Picea rubens*)
River birch (*Betula nigra*)
Rosebay rhododendron (*Rhododendron maximum*)
Round-leaved currant (*Ribes rotundifolium*)
Sandmyrtle (*Leiophyllum buxifolium*)
Sassafras (*Sassafras albidum*)
Scarlet oak (*Quercus coccinea*)
Shagbark hickory (*Carya ovata*)

Shining or Winged sumac (*Rhus capallinum*)
Short-leaf pine (*Pinus echinata*)
Shumard's oak (*Quercus shumardi*)—2003–2004
Silverbell (*Halesia carolina*)
Silver poplar (*Populus alba*) [exotic]
Smooth blackberry (*Rubus canadensis*)
Sourwood (*Oxydendrum arboretum*)
Sessile-leaved bush-honeysuckle (*Diervilla sessilifolia*)
Spicebush (*Lindera benzoin*)
Staghorn sumac (*Rhus hirta*)
Stolon-bearing hawthorn (*Crataegus iracunda*)
Striped maple (*Acer pensylvanicum*)
Sugar maple (*Acer saccharum*)
Sweet birch or Black birch (*Betula lenta*)
Sweetgum (*Liquidambar styraciflua*)
Table-mountain pine (*Pinus pungens*)
Tree-of-heaven (*Ailanthus altissima*) [exotic]
Tuliptree or Yellow-poplar (*Liriodendron tulipifera*)
Umbrella magnolia (*Magnolia tripetala*)
Virginia pine (*Pinus virginiana*)
Yellow buckeye (*Aesculus octandra*)
White ash (*Fraxinus americana*)
White basswood (*Tilia americana* var. *heterophylla*)
White oak (*Quercus alba*)
Wild hydrangea (*Hydrangea arborescens*)
Wild raisin (*Viburnum nudum* var. *cassinoides*)
Winterberry (*Ilex verticiliata*)
Witch-hazel (*Hamamelis virginiana*)
Yellow birch (*Betula lutea*)

APPENDIX III

CHECKLIST OF WILDFLOWERS, HERBS, SEDGES, AND GRASSES REFERENCED IN TEXT

A date next to an entry indicates when the species was discovered as part of recent ATBI research.

Bee balm (*Monarda* sp.)
Blackberry (*Rubus* sp.)
Bluets (*Houstonia* sp.)
Bloodroot (*Sanguinaria canadensis*)
Blue vervain (*Verbena brasiliensis*) [exotic]—2003–2004
Bog blue grass (*Poa paludigena*)—2005
Brazilian watermeal (*Wolffia brasiliensis*)—2003–2004
Broom-corn (*Sorghum bicolor*) [exotic]—2003–2004
Brown-ray knapweed (*Centaurea jacea*) [exotic]—2006
Bulbous wood-rush (*Luzula bulbosa*)—2003–2004
Butterfly-weed (*Asclepias tuberosa*)
Cain's reed-bent grass (*Calamagrostis cainii*)
Cardinal flower (*Lobelia cardinalis*)
Chicory (*Cichorium intybus*)
Chinese yam (*Dioscorea oppositifolia*) [exotic]
Clematis (*Clematis virginiana*)
Coffeeweed (*Senna obtusifolia*)—2003–2004
Columbine (*Aquilegia canadensis*)
Common mullein (*Verbascum thapsus*) [exotic]
Common tansy (*Tanacetum vulgare*) [exotic]—2003–2004
Cone-cup spike-rush (*Eleocharis tuberculosa*)—2005
Creeping phlox (*Phlox stolonifera*)
Crested dwarf iris (*Iris cristata*)
Crowpoison (*Nothoscordum bivalve*)—2003–2004
Cumberland sedge (*Carex cumberlandensis*)—2003–2004
Cutleaf coneflower (*Rudbeckia* sp.)
Daffodil (*Narcissus pseudo-narcissus*)

Doll's-eyes (*Actaea pachypoda*)
Dutchman's-breeches (*Dicentra cucularia*)
Early blue violet (*Viola X palmata*)—2005
Early meadowrue (*Thalictrum dioicum*)
Fibrous root-sedge (*Carex communis* var. *amplisquama*)—2003–2004
Fire pink (*Silene virginica*)
Fly-poison (*Amianthium muscigtoxicum*)
Foamflower (*Tiarella cordifolia*)
Galax (*Galax urceolata*)
Garlic mustard (*Alliaria petiola*) [exotic]
Gattinger's panic grass (P*anicum gattingeri*)—2003–2004
Glomerate sedge (*Carex aggregata*)—2003–2004
Goldenrod (*Solidago* sp.)
Great plantain (*Plantago major*)—2003–2004
Greenbrier (*Smilax* sp.)
Hair grass (*Deschampsia flexuosa*)—2003–2004
Hairy lettuce (*Lactuca hirsuta*)—2003–2004
Heart-leaf aster (*Symphyotrichum cordifolium*)
Heavy sedge (*Carex gravida* var. *lunelliana*)—2003–2004
Hirsute sedge (*Carex complanata*)—2003–2004
Indian-pipe (*Monotropa uniflora*)
Jack-in-the-pulpit (*Arisaema triphyllum*)
Japanese grass (*Bromus japonicus*) [exotic]
Japanese knotweed (*Polygonum cuspidatum*) [exotic]
Joe Pye weed (*Eupatorium maculatum*)
Johnson grass (*Sorghum halepense*) [exotic]
Juniper leaf (*Polypremum procumbens*)–2003–2004
Kral's sedge (*Carex kraliana*)–2003–2004
Large-flowered bellwort (*Uvularia grandiflora*)
Large-flowered trillium (*Trillium grandiflorum*)
Ladies' tresses (*Spiranthes* sp.)
Least duckweed (*Lemna minuta*)—2005
Little brown jug (*Hexastylis arifolia*)
Mercury spurge (*Euphorbia mercurialina*)—2002
Milkweed (*Asclepias* sp.)
Mountain oat grass (*Danthonia compressa*)
Musk thistle (*Carduus nutans*) [exotic]
Orange-eye butterfly-bush (*Buddleja davidii*) [exotic]—2003–2004
Orange jewelweed or Touch-me-not (*Impatiens capensis*)
Orange-red hawkweed (*Hieracium aurantiacum*) [exotic]
Oriental bittersweet (*Celastrus orbiculatus*) [exotic]

Oriental false hawk's-beard (*Youngia japonica*) [exotic]—2003–2004
Painted trillium (*Trillium undulatum*)
Pale jewelweed or Touch-me-not (*Impatiens pallida*)
Pallid violet (*Viola* sp.)
Partridge-berry (*Mitchella repens*)
Pennsylvania blackberry (*Rubus pensilvanicus*)—2006
Periwinkle (*Vinca minor*) [exotic]
Pink turtlehead (*Chelone lyoni*)
Pipsissewa (*Chimaphila maculata*)
Plume grass (*Miscanthus sinensis*) [exotic]
Purple fringed orchid (*Habenaria psycodes*)
Pussy-toes (*Antennaria* sp.)
Ramps or Wild leeks (*Allium tricoccum*, *Allium burdickii*)
Rattlesnake plantain (*Goodyera pubescens*)
Rugel's ragwort or Rugel's Indian plantain (*Rugelia nudicaulis*)
Serrinated skullcap (*Scutellaria serrata*)—2002
Sharp-lobed hepatica or Liverwort (*Hepatica nobilis* var. *acuta*)
Short-beak sedge (*Carex brevior*)—2003–2004
Showy orchis (*Orchis spectabilis*)
Slender crab grass (*Digitaria filiformis*)—2003–2004
Spiny plumeless-thistle (*Carduus acanthoides*)—2006
Spreading avens (*Geum radiatum*)
Spring beauty (*Claytonia virginica*)
Steeplebush (*Spiraea tomentosa*)—2003–2004
Sweet autumn virgin's-bower (*Clematis terniflora*)—2003–2004
Trailing arbutus (*Epigaea repens*)
Trout-lily or Dog-toothed violet or Fawn lily (*Erythronium americanum*)
Turk's cap lily (*Lilium superbum*)
Twinsisters (*Lonicera tatarica*) [exotic]—2005
Velvet leaf blueberry (*Vaccinium myrtilloides*)—2006
Virginia spiraea (*Spiraea virginiana*)
Wakerobin (*Trillium erectum*)
Wand mullein (*Verbascum virgatum*) [exotic]—2003–2004
White-edge sedge (*Carex debilis* var. *debilis*) [new var.]—2006
White fringed phacelia (*Phacelia fimbriata*)
Wild geranium (*Geranium maculatum*)
Wild ginger (*Asarum canadense*)
Wild hydrangea (*Hydrangea arborescens* var. *arborescens*)
Wintergreen or Teaberry or Mountain tea (*Gaultheria procumbens*)
Winter vetch (*Vicia villosa varia*) [new subspecies]—2006
Wirey panic grass (*Panicum flexile*)—2003–2004

Wood sorrel (*Oxalis montana*)
Yellow bead lily (*Clintonia borealis*)
Yellow fringed orchid (*Habenaria ciliaris*)
Yellow lady's-slipper (*Cypripedium parviflorum* var. *pubescens*)
Yellow star grass (*Hypoxis hirsuta*)
Yellow trillium (*Trillium luteum*)

APPENDIX IV

CHECKLIST OF FERNS REFERENCED IN TEXT

A date next to an entry indicates the year the species was discovered as part of recent ATBI research.

Bracken fern (*Pteridium aquilinum*)
Christmas fern (*Polystichum acrostichoides*)
Hay-scented fern (*Dennstaedtia punctilobula*)
Intermediate wood fern (*Dryopteris intermedia*)
Rock polypody (*Polypodium appalachianum*)
Southern lady fern (*Athyrium filix-femina* ssp. *Asplenioides*)
Taylor's filmy fern (*Hymenophyllum tayloriae*)—2003–2004
Walking fern (*Asplenium rhizophyllum*)

Notes

Chapter 1. The Beginning

Cherokee Legend: Mooney, J. 1900. *Myths of the Cherokee.* Nineteenth Annual Report, Bureau of American Ethology. Government Printing Office, Washington, D.C. Reprinted 1996, Dover Publications.

Age of the Earth: Futuyma, D. J. 1986. *Evolutionary Biology.* Second edition. Sinauer Associates, Sunderland, Massachusetts.

Oldest Rocks: Woese, C. R. 1981. Archaebacteria. *Scientific American* 244(6): 106–122; Hayes, J. M. 1996. The earliest memories of life on Earth. *Nature* 384:21–22.

Oldest Living Organisms: Bishop, A. C., A. R. Woolley, and W. R. Hamilton. 2005. *Guide to Minerals, Rocks & Fossils.* Firefly Books Ltd., Buffalo, New York; Allwood, A. C., M. R. Walter, B. S. Kamber, C. P. Marshall, and I. W. Burch. 2006. Stromatolite reef from the Early Archean era of Australia. *Nature* 441:714–718.

Geological Time Scale, Plate Tectonics, Seafloor Spreading: Roberts, D. C. 1996. *A Field Guide to Geology. Eastern North America.* Houghton Mifflin Company, Boston; Luhr, J. F. (ed.). 2003. *Earth.* DK Publishing, New York.

Mountain Building: Moore, H. L. 1988. *A Roadside Guide to the Geology of the Great Smoky Mountains National Park.* The University of Tennessee Press, Knoxville.

Continental Movements: Murphy, J. B., and R. D. Nance. 1992. Mountain belts and the supercontinent cycle. *Scientific American* 266(4):84–91.

Ocoee Rocks: King, P. B., and A. Stupka. 1950. The Great Smoky Mountains—Their Geology and Natural History. *Scientific Monthly,* 71:31–43.

Rock Types: King, P. B., R. B. Neuman, and J. B. Hadley. 1968. Geology of the Great Smoky Mountains National Park, Tennessee and North Carolina. *U. S. Geological Survey Professional Paper* 587. 23 p.

Chapter 2. Topography and Climate

General: SAMAB (Southern Appalachian Man and the Biosphere). 1996. The Southern Appalachian assessment. U.S. Forest Service, Southern Region, Atlanta.

Temperature and Precipitation Data: Weather Summary. GSMNP. December 2000.

Plant Communities: Superintendent's Annual Report—2002.

Monthly Calendar of Natural Events: GSMNP.

Chapter 3. Pre-Park History

Archaeological Excavations: Bass, Q. R. 1977. Prehistoric settlement and subsistence patterns in the Great Smoky Mountains. M.S. thesis, University of Tennessee, Knoxville. 151 p.; Lynch, J. A., and J. S. Clark. 1996. How fire and anthropogenic disturbance shaped forests of the southern Appalachian Mountains USA. *Bulletin of the Ecological Society of America* 77:276; Superintendent's Annual Reports, 2000–2006.

Cades Cove Settlement: Shields, A. R. 1981. *The Cades Cove Story*. Great Smoky Mountains Association, Gatlinburg.

Bison and Elk in Cades Cove: Allen, J. A. 1876. The American Bisons, living and extinct. Memoirs of the Kentucky Geological Survey, Vol. 1, Part 2; Ganier, A. F. 1928. The wild life of Tennessee. *Journal of the Tennessee Academy of Science* 3(3):10–22; Linzey, D. W. 1995. *Mammals of Great Smoky Mountains National Park*. The McDonald & Woodward Publishing Co., Blacksburg, Virginia.

Early Research: Lanman, C. 1849. *Letters from the Alleghany Mountains*. New York: George P. Putnam; Buckley, S. B. 1867. Mountains of North Carolina and Tennessee. *American Journal of Science and Arts*, 2nd series, vol. 27. May 1867; Dunn, E. R. 1920. Some reptiles and amphibians from Virginia, North Carolina, Tennessee and Alabama. *Proceedings of the Biological Society of Washington* 33:129–138; Dunn, E. R. 1926. The salamanders of the family Plethodontidae. Smith College Anniversary Publication, Northampton, Massachusetts. 1–441. Reissued in 1972 by the Society for the Study of Amphibians and Reptiles; Weller, W. H. 1930. A new salamander from the Great Smoky Mountain National Park. *Proceedings of the Junior Society of Natural History, Cincinnati* 1(7):3–4; Weller, W. H. 1931. A preliminary list of the salamanders of the Great Smoky Mountains of North Carolina and Tennessee. *Proceedings of the Junior Society of Natural History, Cincinnati* 2(1):21–32; McClure, G. V. 1931. The Great Smoky Mountains with preliminary notes on the salamanders of Mt. LeConte and LeConte Creek. *Zoologica* 11(6):53–76; Necker, W. L. 1934. Contribution to the herpetology of the Smoky Mountains of Tennessee. Bull. *Chicago Academy of Science* 5:1–4; King, W. 1936. A new salamander (*Desmognathus*) from the Southern Appalachians. *Herpetologica* 1:57–60; King, W. 1939. A survey of the herpetology of Great Smoky Mountains National Park. *American Midland Naturalist* 21:531–582; Komarek, E. V. and R. Komarek. 1938. Mammals of the Great Smoky Mountains. *Bulletin of the Chicago Academy of Sciences* 5(6):137–162; Cole, A. C. 1940. A guide to the ants of the Great Smoky Mountains National Park, Tennessee. *American Midland Naturalist* 24(1):1–88; Stupka, A. 1940. Statement concerning the resignation of Willis King. Superintendent's Monthly Report, November 1940; Thornborough, L. 1942. *The Great Smoky Mountains*. The University of Tennessee Press, Knoxville; Brimley, C. S. 1944. the mammals of North Carolina. 18 installments in Carolina Tips. Carolina Biology Supply Company, Elon College, North Carolina; Neuman, R. B. 1947. Notes on the geology of Cades Cove, Great Smoky Mountains National Park, Tennessee. *Journal of the Tennessee*

Academy of Science 22(3):167–172; Hamnett, W. L. and D. C. Thornton. 1953. Tar Heel wildlife. North Carolina Wildlife Resources Commission, Raleigh, North Carolina; Lix, H. W. 1958. Short history of the Great Smoky Mountains National Park. Typewritten manuscript in GSMNP Library. 126 p.; Linzey, D. W. 1995. Mammals of Great Smoky Mountains National Park. The McDonald & Woodward Publ. Co., Blacksburg, Virginia.

Bounties: Brown, M. L. 2000. *The Wild East.* University Press of Florida, Gainesville (from Codes of Tennessee 1858, 1884).

Logging Disturbance: Kephart, H. 1913. *Our Southern Highlanders.* The Macmillan Company. (Reprinted 1922); Pyle, C. 1985. Vegetation disturbance history of Great Smoky Mountains National Park: An analysis of archival maps and records. USDI National Park Service Research/Resources Management Report SER-77. Great Smoky Mountains National Park, Gatlinburg.

Chapter 4. Park Formation

Efforts to Form Park: Wilson, J. 1901. Report to President Theodore Roosevelt; Longwell, H. C. 1924. Speech. Southern Appalachian National Park Committee meeting in Gatlinburg; Southern Appalachian National Park Committee. Report to Secretary of the Interior. December 12, 1924; Thornborough, L. 1942. *The Great Smoky Mountains.* The University of Tennessee Press, Knoxville; Lix, H. W. 1958. Short history of the Great Smoky Mountains National Park. Typewritten manuscript in GSMNP Library. 126 p.; Campbell, C. C. 1960. *Birth of a National Park in the Great Smoky Mountains.* The University of Tennessee Press, Knoxville.

Walker Family: Madden, R. R., and T. R. Jones. 1977. *Mountain Home.* U.S. Department of the Interior/National Park Service, Washington, D.C.

Acreage of Park, 1940–2008: Data compiled by R. W. Wightman, GSMNP.

Chapter 5. Forests and Balds

Early Studies: Cain, S. A. 1930. Certain floristic affinities of the trees and shrubs of the Great Smoky Mountains and vicinity. *Butler University Botanical Studies* 1:129–150; Cain, S. A. 1935. Ecological studies of the vegetation of the Great Smoky Mountains. II. The quadrant method applied to sampling spruce and fir forest types. *American Midland Naturalist* 16:566–584; Wells, B. W. 1936a. Origin of the Southern Appalachian grass balds. *Science* 83:283; Wells, B. W. 1936b. Andrews Bald: The problem of its origin. *Southern Appalachian Botanical Club Journal (Castanea)* 1:59–62; Wells, B. W. 1937. Southern Appalachian grass balds. *Journal of the Elisha Mitchell Scientific Society* 53:1–26; Cain, S. A. 1943. The tertiary character of the cove hardwood forests of the Great Smoky Mountains National Park. *Torrey Botanical Club Bulletin* 70:213–235; Cain, S. A. et al. 1937. A preliminary guide to the Greenbrier-Brusey [*sic.* Brushy] Mountain nature trail, the Great Smoky Mountains National Park. University of Tennessee, Knoxville. Mimeographed. 29 p.; Russell, N. H. 1953. The beech gaps of the Great Smoky Mountains. *Ecology* 34:366–374.

Park Comparisons, Forest Types: Stupka, A. 1960. *Great Smoky Mountains National Park.* Natural History Handbook Series No. 5, Washington, D.C. 75 p.; Vandermast, D. B.,

D. H. Van Lear, and B. D. Clinton. 2002. American chestnut as an allelopath in the southern Appalachians. *Forest Ecology and Management* 165:173–181.

Vegetation Data: Janet Rock, Botanist, GSMNP (current data); White, P. S. 1982. The flora of Great Smoky Mountains National Park: an annotated checklist of the vascular plants and a review of previous floristic work. *Research/Resource Management Report SER-55.* National Park Service.

Glaciation: Brown, J. H., and A. C. Gibson. 1983. *Biogeography.* C. V. Mosby, St. Louis; Linzey, D. W. 2001. *Vertebrate Biology.* McGraw-Hill, New York.

Southern Range Limits: Linzey, D. W. 1984. Distribution and status of the northern flying squirrel and the northern water shrew in the southern Appalachians. The Southern Appalachian spruce-fir ecosystem: Its biology and threats. National Park Service Research/Resources Management Report SER-71:193–200. Uplands Field Research Laboratory.

Large Trees: Stupka, A. 1964. *Trees, Shrubs, and Woody Vines of Great Smoky Mountains National Park.* University of Tennessee Press, Knoxville.

Old Trees: Manning, R. 1999. *100 Hikes in Great Smoky Mountains National Park.* The Mountaineers, Seattle.

Blue Haze: Kephart, H. 1921. *Our Southern Highlanders.* Macmillan, New York. Reprint, University of Tennessee Press, Knoxville. 1976.

Forest Types: Whittaker, R. H. 1956. Vegetation of the Great Smoky Mountains. *Ecology Monographs* 26(1):1–80; Stevenson, G. B. 1967. *Trees of the Great Smoky Mountains National Park.* The Great Smoky Mountains Natural History Association. 32 p.

Mistletoe (Viscin): Venable, S. 2005. Prodigious pucker power. *Knoxville News Sentinel.* December 11, 2005. B1.

Tallest and Biggest (Largest Girth) Trees: Eastern Native Tree Society, Black Mountain, North Carolina, 2006; Blozan, W. 2006. Personal communication; Hunter, E. 2007. What these trees can do (with a little help from their friends). *Blue Ridge Country* 20 (3–4): 8–9.

National Champion Trees: National Register of Big Trees—2006–2007. American Forests, Washington, D.C.

Coarse Woody Debris: Webster, C. R., and M. A. Jenkins. 2005. Coarse woody debris dynamics in the southern Appalachians as affected by topographic position and anthropogenic disturbance history. *Forest Ecology and Management* 217:319–330.

Changes in Xeric Forests: Harrod, J., P. S. White, and M. E. Harmon. 1998. Changes in xeric forests in western Great Smoky Mountains National Park, 1936–1995. *Castanea* 63(3):346–360.

Heath Balds: Cain, S. A. 1930. An ecological study of the heath balds of the Great Smoky Mountains. Butler University, *Botanical Studies* 1:117–208; Whittaker, R. H. 1956. Vegetation of the Great Smoky Mountains. *Ecological Monographs* 26:1–80; White, P. S., S. P. Wilds, and D. A. Stratton. 2001. The distribution of heath balds in the Great Smoky Mountains, North Carolina and Tennessee, USA. *Journal of Vegetation Science* 12(4):453–466; Conkle, L., and R. S. Young. 2004. New research on heath balds in Great

Smoky Mountains National Park: Surprising results for resource management and for science. Geological Society of America *Abstracts with Programs* 36(5):228.

Grass Balds: Brewer, C. 1993. *Great Smoky Mountains National Park*. Graphic Arts Center Publishing Company, Portland, Oregon; Zeigler, W. G. 1883. *The Heart of the Alleghenies*. A. Williams & Co., Cleveland; Bass, Q. R. 1977. Prehistoric settlement and subsistence patterns in the Great Smoky Mountains. Final Report submitted to National Park Service; Cain, S. A. 1931. Ecological studies of the vegetation of the Great Smoky Mountains of North Carolina and Tennessee. *Botanical Gazette* 91:22–41; Camp, W. H. 1931. The Grass Balds of the Great Smoky Mountains of Tennessee and North Carolina. *Ohio Journal of Science* 31:157–164; Bass, Q. R. 1977. Prehistoric settlement and subsistence patterns in the Great Smoky Mountains. Final Report submitted to National Park Service.

Miscellaneous Data: Superintendent's Annual Reports, 2000–2006.

Chapter 6. Animals

Insect Data: Compiled by Becky Nichols, GSMNP.

Monarch Butterfly: Vertical files, Twin Creeks Science Center. Personal communication with Janice Pelton, June 22, 2006; personal communication with Michelle Prysby, July 24, 2006; interviews with Janice Pelton and Paul Super; data on earliest and latest sightings compiled by Meryl Rose, GSMNP.

Gregorys Cave: Superintendent's Annual Report, 2003; Mays, J. D. 2002. A systematic approach to sampling the arthropod assemblage of Gregorys Cave, Great Smoky Mountains National Park. Master's thesis, Western Carolina University, Cullowhee, North Carolina. 83 p.

Fish: Parker, C. R., and D. W. Pipes. 1990. Watersheds of the Great Smoky Mountains National Park: a Geographical Information System Analysis. *Research/Resource Management Report SER-91/01*. United States Department of Interior, National Park Service; Simbeck, D. J. 1990. *Distribution of the Fishes of the Great Smoky Mountains National Park*. Master's of science thesis, University of Tennessee; Etnier, D. A., and W. C. Starnes. 1993. *The Fishes of Tennessee*. University of Tennessee Press, Knoxville; Moore, S. E., M. A. Kulp, J. A. Hammonds, and B. Rosenlund. 2005. Restoration of Sams Creek and an Assessment of Brook Trout Restoration Methods, Great Smoky Mountains National Park. U.S. Department of Interior, National Park Service. *Technical Report NPS/NRWRD/NRTR-2005/342;* Rakes, P. 2005. Snorkel surveys of Great Smoky Mountains National Park fish. *ATBI Quarterly* 6(2):3; Superintendent's Annual Reports, 2000–2006; Flebbe, P. A., L. D. Roghair, and J. L. Bruggink. 2006. Spatial modeling to project southern Appalachian trout distribution in a warmer climate. *Transactions of the American Fisheries Society* 135:1371–1382; P. Rakes. 2007. Personal communication.

Amphibians: Huheey, J. E., and A. Stupka. 1967. *Amphibians and Reptiles of Great Smoky Mountains National Park*. The University of Tennessee Press, Knoxville; Brodie, E. D., Jr., and R. R. Howard. 1979. Experimental study of Batesian mimicry in the salamanders *Plethodon jordani* and *Desmognathus ochrophaeus*. *American Midland Naturalist* 90:38–46; Licht, L. E. 1991. Habitat selection of *Rana pipiens* and *Rana sylvatica* during exposure to warm and cold temperatures. *American Midland Naturalist* 125:259–268; Costanzo, J. P., R. E. Lee, Jr., and M. F. Wright. 1992. Cooling rate influences cryoprotectant distribution and organ dehydration in freezing wood frogs. *Journal of Experimental Zoology* 261:373–378; Tilley, S. G., and J. E. Huheey. 2001. *Reptiles & Amphibians of the Smokies*. Great Smoky Mountains Natural History Association; Hyde, E. J., and T. R. Simons. 2001. Sampling plethodontid salamanders: sources of variability. *Journal of Wildlife Management* 65(4):624–632; Nickerson, M. A., K. L. Krysko, and R. D. Owen. 2002. Ecological status of the Hellbender (*Cryptobranchus alleganiensis*) and the Mudpuppy (*Necturus maculosus*) salamanders in the Great Smoky Mountains National Park. *Journal of the North Carolina Academy of Science* 118:27–34; Bailey, L. L., T. R. Simons, and K. H. Pollock. 2004. Estimating site occupancy and species detection probability parameters for terrestrial salamanders. *Ecological Applications* 14(3):692–702; Bailey, L. L., T. R. Simons, and K. H. Pollock. 2004. Estimating detection probability parameters for *Plethodon* salamanders using the Robust capture-recapture design. *Journal of Wildlife Management* 68(1):1–13; Bailey, L. L., T. R. Simons, and K. H. Pollock. 2004. Spatial and temporal variation in detection probability of *Plethodon* salamanders using the Robust capture-recapture design. *Journal of Wildlife Management* 68(1):14–24; Cash, B. Range of box turtle. Personal communication, August 21, 2006; Freake, M. Hellbenders. Personal communication, August 22, 2006.

Reptiles: Savage, T. 1967. The diet of rattlesnakes and copperheads in the Great Smoky Mountains National Park. *Copeia* 1967 (1):226–227; Huheey, J. E., and A. Stupka. 1967. *Amphibians and Reptiles of Great Smoky Mountains National Park*. The University of Tennessee Press, Knoxville; Tilley, S. G., and J. E. Huheey. 2001. *Reptiles & Amphibians of the Smokies*. Great Smoky Mountains Natural History Association; Cash, B. Range of box turtle. Personal communication, August 21, 2006.

Birds: Stupka, A. 1963. *Notes on the Birds of Great Smoky Mountains National Park*. The University of Tennessee Press, Knoxville; Friedman, H., and L. F. Kiff. 1985. The parasitic cowbirds and their hosts. *Proceedings of the Western Foundation of Vertebrate Zoology* 2:225–304; Davies, N. B., and M. Brooke. 1991. Coevolution of the cuckoo and its hosts. *Scientific American* 264(1):92–98; Suarez, R. K. 1992. Hummingbird flight: Sustaining the highest mass-specific metabolic rates among vertebrates. *Experientia* 48:565–570; Farnsworth, G. L., and T. R. Simons. 1999. Factors affecting nesting success of wood thrushes in Great Smoky Mountains National Park. *The Auk* 116(4):1075–1082; Simons, T. R., G. L. Farnsworth, and S. A. Shriner. 2000. Evaluating Great Smoky Mountains National Park as a population source for the wood thrush. *Conservation Biology* 14(4):1133–1144; Farnsworth, G. L., and T. R. Simons. 2000. Observations of wood thrush nest predators in a large contiguous forest. *Wilson Bulletin* 112(1):82–87; Shriner, S. A., T. R. Simons, and G. L. Farnsworth. 2002. A GIS-based habitat model for wood thrush, *Hylocichla mustelina*, in Great Smoky Mountains National Park. Chapter 47: 529–535. In: Scott, J. M., P. J. Heglund, M. L. Morrison, J. B. Haufler, M. G. Raphael, W. A. Wall, and F. B. Samson (eds.), *Predicting Species Occurrences: Issues of Accuracy and Scale*. Island Press, Washington, D.C.; Simons, T. R., S. A. Shriner, and G. L. Farnsworth. 2006. Comparison of breeding bird and vegetation communities in primary and secondary forests of Great Smoky Mountains National Park. *Biological Conservation* 129 (2006):302 311.

Bird-Banding Data: Personal communication with Paul Super, June 17, 2006, and August 23, 2006; personal communication with Jeremy Lloyd, June 2006; personal communication with Charlie Muise, August 3, 2006.

Mammals—Shrews: Maynard, C. J. 1889. Singular effects produced by the bite of a short-tailed shrew, *Blarina brevicauda. Contributions to Science* 1:57; Krosch, H. F. 1973. Some effects of the bite of the short-tailed shrew, *Blarina brevicauda. Journal of the Minnesota Academy of Science* 39:21.

Mammals—Coyotes: Resource Management Information Letter. 1982. The coyote—Native or exotic? Great Smoky Mountains National Park. June 23, 1982; Branham, L. 1988. Unstoppable. Man hasn't found way to halt coyote's growth. *Knoxville News-Sentinel.* March 13, 1988.

Mammals—White-tailed Deer: Letters concerning stocking of deer in Park: Eakin to Albright, March 30, 1931; Ganier to Reddington, May 21, 1931; Henderson to Ganier, June 29, 1931; Ganier to Eakin, July 8, 1931; Eakin to Ganier, July 13, 1931; Eakin to Albright, July 13, 1931; Acting Associate Director to Eakin, July 17, 1931—All from the National Archives, Record Group 79, Records of the National Park Service, Entry 7 a, Central Classified Files, 1907–1932, Box 312, National Parks: Great Smoky, Folder 715–04; Webster, C. R., M. A. Jenkins, and J. H. Rock. 2005. Twenty years of forest change in the woodlots of Cades Cove, Great Smoky Mountains National Park. *Journal of the Torrey Botanical Society* 132:280–292; Webster, C. R., M. A. Jenkins, and Janet H. Rock. 2005. Long-term response of spring flora to chronic herbivory and deer exclusion in Great Smoky Mountains National Park, USA. *Biological Conservation* 125:297–307.

Mammals—Beaver: *Park News and Views*, 1966; *Park News and Views*, 1968.

Mammals—Golden Mouse: Linzey, D. W. 1968. An ecological study of the golden mouse, *Ochrotomys nuttalli*, in the Great Smoky Mountains National Park. *American Midland Naturalist* 79(2):320–345; Linzey, D. W. and A. V. Linzey. 1967a. Maturational and seasonal molts in the golden mouse, *Ochrotomys nuttalli. Journal of Mammalogy* 48(2):236–241; Linzey, D. W., and A. V. Linzey. 1967b. Growth and development of the golden mouse, *Ochrotomys nuttalli. Journal of Mammalogy* 48(3):445–458; Linzey, D. W. and R. L. Packard. 1977. *Ochrotomys nuttalli*. Mammalian Species No. 75:1–6.

Mammals—Black Bear: Pelton, M. R., and L. E. Beeman. 1975. A synopsis of population studies of the black bear in the Great Smoky Mountains National Park. *Proceedings of the Southern Regional Workshop of the American Association of Zoological Parks and Aquariums*, Knoxville: 43–48; Hooper, E. 1998. Record black bear poached in park. *Star Journal*, December 8, 1998; Horstman, L. 2001. *The Troublesome Cub in the Great Smoky Mountains.* Great Smoky Mountains Association, Gatlinburg, Tennessee.

Mammals—Bats: Species identification of bats testing positive for rabies: Dr. Lori Sheeler, Epidemiologist, East Tennessee Regional Health Office: personal communication, June 2006.

Mammals—General: Linzey, A. V., and D. W. Linzey. 1971. *Mammals of Great Smoky Mountains National Park.* University of Tennessee Press, Knoxville; Linzey, D. W. 1995. *Mammals of Great Smoky Mountains National Park.* The McDonald & Woodward Publishing Company, Blacksburg, Virginia; Linzey, D. W. 1995. Mammals of Great Smoky

Mountains National Park—1995 Update. *Journal of the Elisha Mitchell Scientific Society* 111(1):1–81; Superintendent's Annual Reports, 2000–2006.

Chapter 7. Endangered Species

Rock Gnome Lichen: Love, J. 2006. Tremont expedition to the wilds of Raven Fork. *ATBI Quarterly* 7(4):1.

Spruce-fir Moss Spider: Coyle, F. A. 1981. The mygalomorph spider genus *Microhexura* (Araneae, Dipluridae). *Bulletin of the American Museum of Natural History* 170:64–75; Fridell, J. A. 1994. Endangered and threatened wildlife and plants; proposal to list the spruce-fir moss spider as an endangered species. *Federal Register* 59(18):3825–3829; Fridell, J. A. Endangered and threatened wildlife and plants; reopening of public comment period and notice of availability of draft economic analysis for proposed critical habitat determination for the spruce-fir moss spider. *Federal Register* 66(29):9806–9808; Coyle, F. A. 2004. Status survey of the endangered spruce-fir moss spider, *Microhexura montivaga*, in the Great Smoky Mountains National Park. Unpublished report to GRSM. 25 p.

Red-cockaded Woodpecker: Fleetwood, R. J. 1936. The red-cockaded woodpecker in Blount County, Tennessee. *Migrant* 7:103; Tanner, J. T. 1965. Red-cockaded woodpecker nesting in the Great Smoky Mountains National Park. *Migrant* 36(3):59; Dimmick, R. W., W. W. Dimmick, and C. Watson. 1980. Red-cockaded woodpeckers in the Great Smoky Mountains National Park: Their Status and Habitat. Research/Resources Management Report No. 38. 21 p.

Northern Flying Squirrel: Linzey, D. W. 1983. Status and distribution of the northern water shrew (*Sorex palustris*) and two subspecies of northern flying squirrel (*Glaucomys sabrinus coloratus* and *Glaucomys sabrinus fuscus*). Final Report to U.S. Fish and Wildlife Service. 42 p.; *Federal Register*. November 21, 1984. Endangered and threatened wildlife and plants: Proposed endangered status for two kinds of northern flying squirrel. *Federal Register* 49(226):45880–45884; U.S. Fish and Wildlife Service. 1990. Appalachian northern flying squirrels (*Glaucomys sabrinus fuscus* and *Glaucomys sabrinus coloratus*) recovery plan. Newton Corner, Mass.; Weigl, P. D., T. W. Knowles, and A. C. Boynton. 1999. The distribution and ecology of the northern flying squirrel, *Glaucomys sabrinus coloratus,* in the Southern Appalachians. North Carolina Wildlife Resources Commission, Raleigh.

Eastern Cougar: Brewer, C. 1964. Hike recalls tales of tall guide, panther wrestling. *Knoxville News-Sentinel*. June 28, 1964; Culbertson, N. 1977. Status and history of the mountain lion in the Great Smoky Mountains National Park. Uplands Field Research Laboratory Research/Resource Management Report No. 15.

General: Linzey, D. W. 1995. *Mammals of Great Smoky Mountains National Park*. The McDonald & Woodward Publishing Company, Blacksburg, Virginia; Linzey, D. W. 1995. Mammals of Great Smoky Mountains National Park—1995 Update. *Journal of the Elisha Mitchell Scientific Society* 111(1):1–81; Superintendent's Annual Reports, 2000–2006.

Chapter 8. Reintroductions

Smoky Madtom and Yellowfin Madtom: Lennon, R. E., and P. S. Parker. 1959. The reclamation of Indian and Abrams Creeks, Great Smoky Mountains National Park. U.S.

Fish and Wildlife Service Special Scientific Report—Fisheries No. 306:1–22; Taylor, W. R. 1969. A revision of the catfish genus *Noturus* Rafinesque with an analysis of higher groups in the Ictaluridae. *Bulletin. United States Natural Museum* 282. 315 p.; Agency Draft Recovery Plan. 1983. Yellowfin Madtom (*Noturus flavipinnis*). Asheville Endangered Species Field Office, U.S. Fish and Wildlife Service. February 1983. 31 p.; Shute, J. R., P. W. Shute, and D. A. Etnier. 1987. Reintroduction of smoky madtom (*Noturus baileyi*) and yellowfin madtom (*N. flavipinnis*) into Abrams Creek, Blount County, Tennessee. 1987 Progress Report. 8 p.; Satterfield, J. 1988. Park Service reintroduces fish species. *Knoxville News-Sentinel*, September 27, 1988; Shute, P. W., J. R. Shute, and D. A. Etnier. 1989. Reintroduction of smoky madtom (*Noturus baileyi*) and yellowfin madtom (*Noturus flavipinnis*) into Abrams Creek, Blount County, Tennessee.1989 Progress Report. 13 p.; Rakes, P. L., P. W. Shute, J. R. Shute, and D. A. Etnier. 1990a. Reintroduction of smoky madtom (*Noturus baileyi*) and yellowfin madtom (*Noturus flavipinnis*) into Abrams Creek, Blount County, Tennessee. 14 p.; Rakes, P. L., P. W. Shute, J. R. Shute, and D. A. Etnier. 1990b. Reintroduction of smoky madtom (*Noturus baileyi*) and yellowfin madtom (*Noturus flavipinnis*) into Abrams Creek, Blount County, Tennessee. 5 p.; Restoration Plan Approval and Finding of No Significant Impact. Smoky and Yellowfin Madtom Restoration Program. Great Smoky Mountains National Park. 5 p.; Rakes, P. L., P. W. Shute, and J. R. Shute. 1992. Reintroduction of smoky madtom (*Noturus baileyi*) and yellowfin madtom (*Noturus flavipinnis*) into Abrams Creek, Blount County, Tennessee. 1992 Progress and Populations Status Report. 13 p.; Shute, J. R., P. L. Rakes, and P. W. Shute. 2005. Reintroduction of four imperiled fishes in Abrams Creek, Tennessee. *Southeastern Naturalist* 4(1):93–110.

Spotfin Chub: Moore, S. E. 1988. Memorandum: Spotfin Chub Reintroduction to Abrams Creek. November 16, 1988. 4 p.

Duskytail Darter: Jenkins, R. C. 1976. A list of undescribed freshwater fish species of continental United States and Canada, with additions to the 1970 checklist. *Copeia*: 642–644; Jenkins, R. C., and N. M. Burkhead. 1994. *The Freshwater Fishes of Virginia*. American Fisheries Society, Bethesda, Maryland; Lennon, R. E., and P. S. Parker. 1959. The reclamation of Indian and Abrams creeks, Great Smoky Mountains National Park. U.S. Fish and Wildlife Service Special Science Report—Fish. 306; Simbeck, D. J. 1990. Distribution of the fishes of the Great Smoky Mountains National Park. Master's thesis, University of Tennessee, Knoxville.

River Otter: Griess, J. M. 1987. *River otter reintroduction in Great Smoky Mountains National Park*. Master's thesis, University of Tennessee, Knoxville; Miller, M. C. 1992. *Reintroduction of river otters into Great Smoky Mountains National Park*. Master's thesis, University of Tennessee, Knoxville.

Elk: GSMNP Elk Progress Reports 1–33, February 2001–July 2006; Simmons, M. 2006. Park sets elk calves record. *Knoxville News-Sentinel*, July 10, 2006. B1, 7.

Red Wolf: Linzey, D. W. 1995. *Mammals of Great Smoky Mountains National Park*. The McDonald & Woodward Publishing Company, Blacksburg, Virginia. 140 p.

Peregrine Falcon: Stupka, A. 1963. *Notes on the Birds of Great Smoky Mountains National Park*. University of Tennessee Press, Knoxville; Knight, R., and L. Benton. Undated. Typewritten untitled report concerning release of peregrine falcons in Park. In vertical files, GSMNP Library; Knight, R. L., and R. L. Shumate. 1984. Typewritten untitled report concerning release of peregrine falcons in Park. In vertical files, GSMNP Library; Knight, R. L., and R. M. Hatcher. 1997. Recovery efforts result in returned nesting of peregrine falcons in Tennessee. *Migrant* 68(2):33–39; 2002–2006 data, personal communication from Paul Super, August 10, 2006.

General: Superintendent's Annual Reports, 2000–2006.

Chapter 9. Natural History Research in the Park

Nature Journal: Stupka, A. 1935–1962. Nature journal, Great Smoky Mountains National Park. 28 volumes (years) each with index. In Great Smoky Mountains National Park library.

Early Research: Whittaker, R. H. 1952. A study of summer foliage insect communities in the Great Smoky Mountains. *Ecological Monographs* 22:1–44; Cole, A. C. 1953. A checklist of the ants (Hymenoptera: Formicidae) of the Great Smoky Mountains National Park, Tennessee. *Journal of the Tennessee Academy of Science* 28(1):34–35.

Invertebrates Named for Arthur Stupka: Data supplied by Annette Hartigan, Librarian, GSMNP.

Don De Foe Biographical Data: Supplied by Annette Hartigan, Librarian, GSMNP.

Susan Bratton: Lange, L. 1976. Ivy Leaguer counts park flowers. *Knoxville News-Sentinel*, September 14, 1976:10; Vertical files, GSMNP.

Uplands Field Research Laboratory: Brochure. Vertical files, GSMNP.

Chapter 10. All Taxa Biodiversity Inventory

DLIA Information: Supplied by Jeannie Hilten, coordinator.

Taxa Table: Prepared by Becky Nichols, co-coordinator, ATBI, GSMNP.

Diatoms: Makosky, S. 2004. Diatoms? Why Diatoms? *ATBI Quarterly*, Spring, 5(2):6.

Tardigrades: Bartels, P. 2005. "Little known" water bears. *ATBI Quarterly*, Spring; Bartels, P. J., and D. R. Nelson. 2005. Tardigrade inventory status report. Annual ATBI Conference, Gatlinburg, Tenn.; Bartels, P. J., and D. R. Nelson. 2006. A large-scale, multi-habitat inventory of the Phylum Tardigrada in the Great Smoky Mountains National Park, USA: a preliminary report. *Hydrobiologia* 558:111–118; Nelson, D. R. and P. J. Bartels. 2007. "Smoky Bears"—Tardigrades of the Great Smoky Mountains National Park. *American Midland Naturalist* (in press); Bartels, P. J., and D. R. Nelson. 2007. An evaluation of species richness estimators for tardigrades of the Great Smoky Mountains National Park, Tennessee and North Carolina, USA. *Journal of Limnology* (in press). New vascular plants 2002–2006: Data compiled by Janet Rock and Becky Nichols, GSMNP. July 2006; Velvet leaf blueberry, *Asheville Citizen Times*, June 29, 2006.

Chapter 11. Environmental Concerns

Air Quality: Finkelstein, P. L., A. W. Davison, H. S. Neufeld, T. P. Meyers, and A. H. Chappelka. 2004. Subcanopy deposition of ozone in a stand of cut leaf coneflower. *Envi-

ronmental Pollution 131:295–303; Stephens, J. 2004. Smokies Named to List of Endangered Parks. Air Pollution, Inadequate Funding Returns Popular National Park to the List. http://www.npca.org/media_center/smokies.asp. Superintendent's Annual Reports, 2000–2005.

Water Quality: Superintendent's Annual Reports, 2000–2006.

Exotic Insects and Diseases: Kuykendall, N. W., III. 1978. Composition and structure of replacement forest stands following southern pine beetle infestations as related to selected site variables in the Great Smoky Mountains. M.S. thesis, University of Tennessee, Knoxville. 122 p.; Harmon, M. E. 1980. Influence of fire and site factors on vegetation pattern and process: a case study of the western portion of Great Smoky Mountains National Park. M.S. thesis, University of Tennessee, Knoxville. 170 p.; Nicholas, N. S., and P. S. White. 1984. The effect of southern pine beetle on fuel loading in yellow pine forest of Great Smoky Mountains National Park. National Park Service/Research/Resource Management Report SER-73. 31 p.; Farnsworth, G. L., and T. R. Simons. 1999. Factors af-fecting nesting success of wood thrushes in Great Smoky Mountains National Park. *Auk* 116(4):1075–1082; Jenkins, M. A., and P. S. White. 2002 *Cornus florida* L. mortality and understory composition changes in western Great Smoky Mountains National Park. *Journal of the Torrey Botanical Society* 129(3):194–206; Simmons, M. 2003. American chestnuts: saving the seed of forest giants. *Knoxville News Sentinel*, November 10, 2003; Straub, B. 2005. Chestnut tree's return likely won't come easy. *Knoxville News Sentinel*, November 25, 2005; Johnson, K. 2006. Summary of Forest Insect and Disease Impacts. Great Smoky Mountains National Park Briefing Statement. January 2006; Holzmueller, E., S. Jose, M. A. Jenkins, A. E.. Camp, and A. J. Long (in press). Dogwood anthracnose: what is known and what can be done? *Journal of Forestry*; Superintendent's Annual Re-ports, 2000–2006; Holzmueller, E. J. 2006. Ecology of flowering dogwood (*Cornus florida* L.) in response to anthracnose and fire in Great Smoky Mountains National Park, USA. Ph.D. dissertation. University of Florida, Gainesville; Holzmueller, E. J., S. Jose, and M. A. Jenkins. 2007. In-fluence of calcium, potassium, and magnesium on *Cornus florida* L. density and resistance to dogwood anthracnose. *Plant and Soil* 290:189–199; Hunter, E. 2007. What these trees can do (with a little help from their friends). *Blue Ridge Country*. March/April 2007:8–9; Jenkins, M. A., S. Jose, and P. S. White. (in press) Impacts of an exotic fungal disease and changes in vegetation communities on foliar calcium cycling in Appalachian forests. *Ecological Applications*; Holzmueller, E. J., S. Jose, and M. A. Jenkins. (in press). Influence of *Cornus florida* L. on calcium mineralization in two southern Appalachian forest types. *Forest Ecology and Management*; Save Our Hemlocks from Hemlock Woolly Adelgid (www.saveourhemlocks.com).

Wild Boar: Stegeman, L. J. 1938. The European wild boar in the Cherokee National Forest, Tennessee. *Journal of Mammalogy* 19:279–290; Bratton, S. P. 1974. The effect of European wild (*Sus scrofa*) on the high-elevation vernal flora in Great Smoky Mountains National Park. *Bulletin of the Torrey Botanical Club* 101:198–206; Bratton, S. P. 1975. The effect of the European wild boar (*Sus scrofa*) on the Gray Beech Forest in the Great Smoky Mountains. *Ecology* 56:1356–1366; Belden, R. C., and M. R. Pelton. 1975. European wild hog rooting in the mountains of East Tennessee. *Proceedings of the Annual Conference of the Southeastern Association of Game and Fish Commissioners* 29:665–671; Scott, C. D.

and M. R. Pelton. 1975. Seasonal food habits of the European wild hog in the Great Smoky Mountains National Park. *Proceedings of the Annual Conference of the Southeastern Association of Game and Fish Commissioners* 29:585–593; Belden, R. C., and M. R. Pelton. 1976. Wallows of the European wild hog in the mountains of east Tennessee. *Journal of the Tennessee Academy of Science* 51:91–93; Howe, T. D., and S. P. Bratton. 1976. Winter rooting activity of the European wild boar in the Great Smoky Mountains National Park. *Castanea* 41:256–264; Huff, M. H. 1977. The effect of European wild boar (*Sus scrofa*) on the woody vegetation of gray beech forest in the Great Smoky Mountains. Management Report No. 18, Uplands Field Research Laboratory, Great Smoky Mountains National Park, Gatlinburg, Tennessee; Ackerman, B. B., M. E. Harmon, and F. J. Singer. 1978. Studies of the European wild boar in the Great Smoky Mountains National Park. First Annual Report, Part II: Seasonal food habits of European wild boar, 1977. Uplands Field Research Laboratory, Great Smoky Mountains National Park, Gatlinburg, Tennessee; Singer, F. J., and B. B. Ackerman. 1981. Food availability, reproduction, and condition of European wild boar in Great Smoky Mountains National Park. *Research/Resources Management Report No. 43.* National Park Service; Singer, F. J., W. T. Swank, and E. E. C. Clebsch. 1982. Some ecosystem responses to European wild boar rooting in a deciduous forest. *Research/Resources Management Report No. 54*:1–31. National Park Service; Bratton, S. P., M. E. Harmon, and P. S. White. 1982. Patterns of European wild boar rooting in the Western Great Smoky Mountains. *Castanea* 47:230–242; Johnson, K. G., R. W. Duncan, and M. R. Pelton. 1982. Reproductive biology of European wild hogs in the Great Smoky Mountains National Park. *Proceedings of the Annual Conference of the Southeastern Association of Fish and Wildlife Agencies* 36:552–564; Lacki, M. J., and R. A. Lancia. 1986. Effects of wild pigs on beech growth in Great Smoky Mountains National Park. *Journal of Wildlife Management* 50:655–659; Peine, J. D., and J. A. Farmer. 1990. Wild hog management at Great Smoky Mountains National Park. *Proceedings of the Verterbrate Pest Conference* 14:221–227; New, J. C., Jr., K. DeLozier, C. E. Barton, P. J. Morris, and L. N. D. Potgieter. 1994. A serologic survey of selected viral and bacterial diseases of European wild hogs, Great Smoky Mountains National Park, USA. *Journal of Wildlife Diseases* 30:103–106; Linzey, D. W. 1995. Mammals of Great Smoky Mountains National Park—1995 Update. *Journal of the Elisha Mitchell Scientific Society* 111:1–81; Diderrich, V., J. C. New, G. P. Noblet, and S. Patton. 1996. Serologic survey for *Toxoplasma gondii* antibodies in free-ranging wild hogs (*Sus scrofa*) from the Great Smoky Mountains National Park and from sites in South Carolina. *Journal of Eukaryotic Microbiology* 43:122; Stiver, W. H., and E. K. DeLozier. 2005. Great Smoky Mountains National Park wild hog control program. Paper presented at Wild Pig Symposium (in press).

Exotic Plants: Great Smoky Mountains National Park Management Folio #4. 2005. Great Smoky Mountains Association, Gatlinburg, Tennessee. 4 p.

Chapter 12. What the Future May Hold

Global Warming: Miller, G. T., Jr. 2000. *Sustaining the Earth*. 4th Edition. Brooks/Cole Publishing Company, Pacific Grove, California; Both, C., and M. E. Visser. 2001. Adjustment to climate change is constrained by arrival date in a long-distance migrant bird. *Nature* 411:296–298; Gian-Reto, W., E. Post, P. Convey, A. Menzel, C. Parmesan.

T. J. C. Beebee, J-M Fromentin, O. Hoegh-Guldberg, and F. Bairlein. 2002. Ecological responses to recent climate change. *Nature* 416:389–395; Parmesan, C., and G. Yohe. 2003. A globally coherent fingerprint of climate change impacts across natural systems. *Nature* 421:37–42; IPCC (eds). *Climate Change 2005. The Scientific Basis. Contribution of Working Group to the Third Assessment Report of the Intergovernmental Panel on Climate Change.* Cambridge University Press, Cambridge; *NOAA Magazine.* 2006. U.S. experienced record warm first half of year, widespread drought and northeast record rainfall. July 14, 2006. http://www.noaanews.noaa.gov/stories2006/s2663.htm; Mohan, J. E., L. H. Ziska, W. H. Schlesinger, R. B. Thomas, R. C. Sicher, K. George, and J. S. Clark. 2006. Biomass and toxicity responses of poison ivy (*Toxicodendron radicans*) to elevated atmospheric CO_2. *Proceedings of the National Academy of Sciences* 103(24):9086–9089: Gagnon, D., J. Rock, and P. Nantel. 2005. Wild American ginseng populations in the southern Appalachians may be negatively affected by climate change. Paper presented at 90th annual conference of the Ecological Society of America, Montreal; Bjerklie, D., Global warming. *Time* 167(14): 27–42; Overpeck, J. T., B. L. Otto-Bliesner, G. H. Miller, D. R. Muhs, R. B. Alley, and J. T. Kiehl. 2006. Paleoclimatic evidence for future ice-sheet instability and rapid sea-level rise. *Science* 311(5768):1747–1750; Hansen, J., M. Sato, R. Ruedy, K. Lo, D. Lea, and M. Medina-Elizade. 2006. Global temperature change. *Proceedings of the National Academy of Sciences* 103(39):14288–14293; National Parks and Conservation Association: (http://www.npca.org/media_center/smokies.asp) and (http://www.npca.org/across_the_ nation/visitor_ experience/code_red); IPCC Report. 2007. *The Physical Science Basis: A Summary for Policymakers.*

Selected Bibliography

Alsop, F. J., III. 1995. Birds of the Great Smoky Mountains [checklist]. Great Smoky Mountains Natural History Association, Gatlinburg, Tennessee.

———. 2003. *Birds of the Smokies.* Great Smoky Mountains Association, Gatlinburg, Tennessee.

Ayers, H., J. Hager, and C. E. Little (editors). 1998. *An Appalachian Tragedy: Air Pollution and Tree Death in the Eastern Forests of North America.* Sierra Club Books, San Francisco.

Bentley, S. L. 2000. *Native Orchids of the Southern Appalachian Mountains.* University of North Carolina Press, Chapel Hill.

Brodo, I. M., S. D. Sharnoff, and S. Sharnoff. 2001. *Lichens of North America.* Yale University Press, New Haven, Connecticut.

Chiles, M. R. 1978. Geographic Place Names in the Great Smoky Mountains National Park. Typewritten manuscript. In GSMNP library.

De Foe, D., K. R. Langdon, and J. Rock. 1989. *Flowering plants of the Great Smoky Mountains National Park—revised 1995.* National Park Service.

Dodd, C. K., Jr. 2004. *The Amphibians of Great Smoky Mountains National Park.* University of Tennessee Press, Knoxville.

Etnier, D. A., and W. C. Starnes. 1993. *The Fishes of Tennessee.* University of Tennessee Press, Knoxville.

Evans, M. 2005. *Ferns of the Smokies.* Great Smoky Mountains Association, Gatlinburg, Tennessee.

Frome, M. 1980. *Strangers in High Places.* Revised ed. University of Tennessee Press, Knoxville.

Gupton, O. W., and F. C. Swope. 1987. *Fall Wildflowers of the Blue Ridge and Great Smoky Mountains.* University Press of Virginia, Charlottesville.

Hoffman, H. L. 1964. Checklist of vascular plants of the Great Smoky Mountains. *Castanea* 29:1–45.

———. 1966. Supplement to checklist, vascular plants, Great Smoky Mountains. *Castanea* 31(4):307–310.

Houk, R. 1993. *A Natural History Guide—Great Smoky Mountains National Park.* Houghton Mifflin Company, Boston.

———. *The Walker Sisters of Little Greenbrier.* Great Smoky Mountains Association, Gatlinburg, Tennessee.

Hutchins, R. E. 1971. *Hidden Valley of the Smokies: With a Naturalist in the Great Smoky Mountains.* Dodd, Mead and Company, New York.

Kemp, S. 2006. *Trees of the Smokies.* Great Smoky Mountains Association, Gatlinburg, Tennessee.

Lennon, R. E. 1960. *Fishes of Great Smoky Mountains National Park.* U.S. Fish and Wildlife Service.

———. 1961. An annotated list of the fishes of Great Smoky Mountains National Park. Fish Control Laboratory, La Crosse, Wisconsin.

Myers, B. T. 2004. *The Walker Sisters: Spirited Women of the Smokies.* Myers & Myers, Publishers, Maryville, Tennessee.

National Park Service. 1981. *Great Smoky Mountains.* Handbook 112. U.S. Department of the Interior, Washington, D.C.

Pierce, D. S. 2000. *The Great Smokies: From Natural Habitat to National Park.* University of Tennessee Press, Knoxville.

Schenck, M. J. 1938. An annotated index to salamanders in Willis King's field notes, Great Smoky Mountains National Park, 1935–1938.

Schmidt, R. G., and W. S. Hooks. 1994. *Whistle over the Mountains: Timber, Track, and Trails in the Tennessee Smokies.* Graphicom Press, Inc., Yellow Springs, Ohio.

Stevenson, G. B. 1985. *Birds of Great Smoky Mountains National Park.* Self-published by author.

Stupka, A. 1965. *Wildflowers in Color.* Harper and Row, New York.

White, P., T. Condon, J. Rock, C. A. McCormick, P. Beaty, and K. Langdon. 2003. *Wildflowers of the Smokies.* Great Smoky Mountains Association, Gatlinburg, Tennessee.

White, P. S., and B. E. Wofford. Rare native Tennessee vascular plants in the flora of Great Smoky Mountains National Park. *Journal of the Tennessee Academy of Science* 59(3):61–64.

INDEX

Page numbers in **boldface** refer to illustrations.

A

Abrams Creek, 22, 107, 111, 115, 147, 172, **173**, 174, 175, 176
Abrams Falls, 123, **173**
Abrell, Joe, 193
acid deposition, 65, 207–8
acid precipitation, 19, 20
acid rain, 208, 209, 211
acidification, 209
air pollution, 207–11
    sulfates, 208
air quality, 208–11
    monitoring station–Look Rock, **209**
Alarka Creek, 147
Albright Grove, 70, 227
ALCOA Power Generating Inc., 48
algae
    number of in park, 55
    diatoms, **201**–2
    green, **201**
All Taxa Biodiversity Inventory, 183, 187, 192, 194, 197–205
Alleghany Mountains, 51
Alligator River National Wildlife Refuge, 180
Alum Cave Bluff, 20, 90, 182
Alum Cave Creek, 89
Alva, Herman, 186
American Chestnut Foundation (TACF), 219

amphibians, 111–21
    species of in park, 112–13, 117–18
    See also frogs; salamanders; toads
Andes Mountains, 11
Anthony Creek, 72
Appalachian Bear Center (ABC), 157
Appalachian Highlands Science Learning Center, 240–41
Appalachian Mountains, 11, 14
    formation of, 8, 11
Appalachian Trail, 14, 90, 167, 225
arthropods, 104
    See also centipedes; invertebrates (aquatic); insects; millipedes; spruce-fir moss spider
Ash Camp Branch, 110
Atlantic Ocean
    formation of, 11

B

bacteria, 99, 103
    Bacillus thuringiensis, 102
Bailey, Larissa, 111
balds, 88–95
    origin of, 90–91
    grass, 34, 91–95, **92**, 232
        Andrews, 91, **94**, 188, 218
        Gregorys, **17**, 90, 91, 92, 93, 94, 187, 218, 231
        Spence Field, **95**, 167
        Silers, 91
    heath, 24, 88–91
Balsam Mountain Road, 167, 215

Balsam Mountain, 62, 166, 179, 215
Barrows, W. M., 37
Bartels, Paul, 203
Bartram, William, 36
Baskins Creek Falls, **15**, 77
Bass, Q. R., 92
Batesian mimicry, 115
Bear Creek, 110
Beech Flats Prong, 95
Beech gap, 231
beechdrops, 67
Big Cataloochee, 63
Big Creek, 177
Big Fork Ridge, 86
birds, 129
    species of in park, 130–35
    American crow, 135
    American goldfinch, 29, 135
    banding of, 140–41
    barn owl, 182
    barred owl, 29
    belted kingfisher, 29, **136**
    black and white warbler, 29
    blackburnian warbler, 29, 226
    black-capped chickadee, 29, 65
    black-throated green warbler, 226
    bluebird, 186
    blue-headed vireo, 226
    bobwhite quail, 186, 235
    broad-winged hawk, 29
    brown creeper, 65
    brown-headed cowbird, 136–**37**
    Canada warbler, 29
    Carolina chickadee, 65, 239
    Carolina wren, 186
    cardinal, 186
    cedar waxwing, 29
    chestnut-collared longspur, 198
    chestnut-sided warbler, 29, 137, 141
    common raven, 65
    crow, 35, 65
    dark-eyed junco, 135–36, 141, 239
    downy woodpecker, 186
    English sparrow, 186
    evening grosbeak, 29
    flicker, 186
    golden-crowned kinglet, 65
    goldfinch, 186
    great horned owl, 29
    harlequin duck, 198
    indigo bunting, **131**
    killdeer, 186
    kingfisher, 186
    Louisiana waterthrush, 29, 138, 140
    mockingbird, 186
    mourning warbler, 198
    nesting of, 135
    phoebe, 186
    pigeon, 35
    pileated woodpecker, **133**, 235
    pine siskin, 29
    peregrine falcon, 161, 162, 164, 172, **181**–82
    purple finch, 29
    red-breasted nuthatch, 29, 65, 138
    red-cockaded woodpecker, 162, **164**–65
    redhead duck, 198
    ruby-throated hummingbird, 29, 138–40, **139**, 141
        wing movements, **139**
    ruffed grouse, 29, 36, 180, 230
    saw-whet owl, 65
    scarlet tanager, 29
    screech owl, 65
    short-billed dowitcher, 198
    short-eared owl, 198
    solitary vireo, 29
    starling, 186
    tufted titmouse, 135
    veery, 29, 65
    waxwing, 186
    white-breasted nuthatch, 138, **139**
    white-throated sparrow, 186
    wild turkey, 29, 36, 38, 80, **138**, 169
    winter wren, 29, 65, 186
    wood thrush, 29, 65, 137, 140, 226
    woodpecker, 236
    yellow-rumped warbler, 29
    yellow-throated vireo, 29
    yellow-throated warbler, 29
Blanket Mountain, 166
Blozan, Will, 59, 70, 73, 82, 84, 86

Blue Ridge Mountains, 14
Blue Ridge Province, 11
Boogerman Loop Trail, 227
Boogerman Pine, 84
Bratton, Susan, 192–93
bryophytes
  hornworts, 53
  liverworts, 53
  mosses, 53
Bt corn, 102
Buckley, S. B., 36

C

Cades Cove, 8, **17**, **28**, 33, 34, 35, 40, 88, 101, 102, 111, 114, 117, 119, 120, 127, 146, 148, 149, 158–59, 168, 169, 180, 182, 187, 203, 210, 211, 226, 227, 230, 231
Cain, S. A., 53
Cain, Stanley, 92, 187
calcium, 220, 221, 222, 233
calderwood, 230
Caldwell, James, 34
Caldwell, Levi, 34
Calle, Fred, 187
Camp, H. W., 37
Camp, W. H., 53, 92
carbon, 209
carbon dioxide, 236–37
Cash, Ben, 122
Cataloochee, 34, 35, **46**, 71, 86, 146, 169, 176, 177, 179, 227
caves
  Blowhole, 18, 165
  Bull, 17, 165
  Gregory, **18**
  Rainbow, **18**
  Scott, **165**
centipedes, 97, **98**
Chambers Creek, 177
Champion Fibre Company, 37, **42**
Charlies Bunion, 14, 62, 188
checklists
  ferns, 261
  trees, shrubs, and vines, 253–56
  wildflowers, herbs, sedges, and grasses, 257–60

Cherokee Indian Reservation. *See* Qualla Boundary
Cherokee National Forest, 159, 174
Cherokee Orchard, 47, 111, 151
Cherokees, 48
  history of, 31–34
chestnuts, 38
Chilhowee Lake, 172
Chilhowee Reservoir, 173
Chilhowee, 111
Chilhowie Mountain, 8, 17, 21, 102
Chimney Tops, 14, **15**, 19
Chimneys campground, 169
Chimneys picnic area, 115, 169, 182
Christmas Bird Count, 192
Citico Creek, 174
Clingmans Dome, 14, 18, 22, 23, 29, 61, 62, 115, 163, 208, 209, 210, 215
  road, 115, 166
coarse woody debris, 58
Cocke County, Tennessee, 179
Cole, A. C., 187
Cole, A. H., 186
Coolidge, President Calvin, 49
copper, 233
Cosby, 88, 127, 151, 179, 227
  campground, 27, 128, 182, 213
  Knob, 62
  ranger station, 182
Cove Mountain, 27, 210
Coyle, Fred, 163
Culbertson, N., 167–68, 193

D

Dalton Gap, 165
Davenport Gap, 91, 168
Davenport, William, 91
Davis, N. S., 37
De Foe, Don, 183, **191**–92
de Soto, Hernando, 32
Deep Creek, 115, 147, 151, 177
DeLozier, Kim, 157
DeWaard, Wanda, 101
diatoms, 201–2
Digelius, Gunnar, 186
Diplopoda, 97

Discover Life in America, Inc., 198
diseases
- beech bark disease (BBD), 224
    - *Cryptococcus fagisuga*, 224
    - *Nectria* fungus, 224
- butternut canker, 229
- *Discula destructiva*, 219
- chestnut blight (*Cryphonectria parasitica*), 216–19
- dogwood anthracnose, 219–22
- dutch elm disease (DED), 219
- elm yellows, 219

DLIA. *See* Discover Life in America, Inc
Dodd, Ken, 121
Double Springs Gap, 62
Dreisback, R. R., 187
Dunn, Emmet Reed, 36
Dunn's Creek, 227

E

Eager, Christopher, 193
Eagle Creek, 147, 177
Eakin, J. Ross, 49, 147, 148
Earth
- age of, 6
- Bering land bridge, 11
- formation of, 5
- Gondwana, 8
- Laurasia, 8
- Laurentia, 8
- Pangaea, 8, 11
- plate tectonics, 7, 8, 13, 15, 18

Eastern Native Tree Society, 59, 71, 82, 84
Elkmont, 37, 38, **39**, 99, 157, 210, 224
Ellis, Vincent, 192, 193
endangered species, 161–69
- Carolina northern flying squirrel, 162, 165–67
- duskytail darter, 162, 164, 175–76
- eastern cougar, 162, 167–69
- gray wolf, 162
- Indiana bat, 162, 165
- red-cockaded woodpecker, 162, 164–65
- red wolf, 162, 167, 179–81
- rock gnome lichen, 161, 162
- smoky madtom, 162, 164, 172–74
- spreading avens, 161–62
- spruce-fir moss spider, 162, 163–64
- yellowfin madtom, 162, 164, 172–74

Endangered Species Act of 1973, 163, 174
epochs
- Eocene, 9
- Paleocene, 9
- Miocene, 9
- Oligocene, 9
- Pleistocene, 9
- Pliocene, 9
- Recent (Holocene), 9

erosion, 18, 35, 38
Evison, Boyd, 193
exotic plants, 233–34

F

Farnsworth, George, 137, 226
fault, 13, 15
- Gatlinburg, 16
- Great Smoky, 16, 17
- Greenbrier, 16
- Oconaluftee, 16, 17
- Thrust, 15
    - Great Smoky Overthrust, 16

ferns, 53
- checklist, 261
- christmas fern, **80**, 81
- common polypody, 82
- eastern bracken, 79
- fiddleheads, **81**
- hayscented, 64
- lady, 64
- polypody, 63, 64
- walking, **77**
- woodfern, 64

Fighting Creek Gap, 16
Finley, Nancy, 194
fire, 34, 38, 58, 213
- prescribed burning, 87–88

fish
- species of in park, 105–07
- brook trout, 38, 108–10, **109**
- brown trout, 108–10

duskytail darter, 162, 164, 174, **175**–76
electroshocking of, 107–**8**
rainbow trout, 108–10, **109**, 193
rock bass, 111
smallmouth bass, 111
smoky madtom, 162, 164, **172**, **173**
spotfin chub, 162, 164, 174, 175
trout, 36, 108–10
yellowfin madtom, 162, 164
Fleetwood, R. J., 164
flowers
    color production in, 26–27
Fontana Dam, 48
Fontana Lake, 48, 231
Fontana Reservoir, 147
Fontana Village, 167
Foothills Parkway, 48
forest types, 52, 59
    beech, 224
    cove hardwood forest, 59, 69–78, **71**, 220
    hemlock forest, 59, 81–82
    northern hardwood forest, 59, 66–69
    oak-hickory forest, 221, 222
    pine-oak forest, 59, 78–81, 220, 221, 222
    spruce-fir forest, 59, **61**–**65**, 73, 88, 116, 163, 208, 239
Forge Creek Road, 165
Fork Ridge, 225
Forney Creek, 177
Forney Ridge, 94, 225
fossils
    brachiopods, 7, 17
    trilobites, 7, 9, 17
Freake, Michael, 115
Friends of Great Smoky Mountains National Park, 198, 243
frogs and toads
    species of in park, 117–18
    Bufonidae (toads), 118
        American, 117–18
        Fowler's, 118
    Hylidae (treefrogs), 119
        northern cricket frog (*Acris crepitans*), 111
        chorus frog, 29
        gray treefrog, 119
        spring peeper, 29, 119
        upland chorus, 119
    Microhylidae, 119
        narrow-mouthed toad, 119
    Pelobatidae, 120
        eastern spadefoot toad, 120, 198
    Ranidae, 119
        bullfrog, 119
        eastern wood frog, 29, 119–20
        green frog, 118, 119
        leopard frog, 119
        pickerel frog, 119
fruits
    apples, 27
    blueberries, 27
    grapes, 27
    plum, 59
    color production in, 27
fungi, 78, 99, 103
    number of in park, 55
    bird's nest, **78**
    bracket, 78
    caesar's amanita mushroom, **55**
    coral, 78
    earthstars, 78
    fluted bird's-nest (*Cyathus striatus*), 78
    morels, 78
    Nectria, 224
    stem canker (*Siroccus clavigigenentijugulandacearum*) 229
    turkey tail, **54**

G

Gastropoda, 98–99
Gatlinburg, 23, 27, 227
geologic eras
    Archean, 7, 9, 13
    Cenozoic, 7, 8, 9
    Mesozoic, 7, 8, 9
    Paleozoic, 7, 8, 9, 17
    Proterozoic, 7, 9
geologic periods
    Cambrian, 7, 8, 9, 16

geologic periods (cont.)
    Carboniferous, 7, 8, 9
    Cretaceous, 8, 9, 11
    Devonian, 7, 9
    Jurassic, 8, 9, 11
    Mississippian, 9
    Ordovician, 7, 9
    Pennsylvanian, 9
    Permian, 7, 9, 10
    Silurian, 7, 8, 9
    Triassic, 8, 9, 10, 11
Giardia lamblia, 212
Giardiasis, 212
glaciation
    effects of, 55–57
global warming, 116, 236–39
Gourley Pond, 102, 120
Grannan, Ted, **104**
grass
    Andropogon, 185
    Japanese, 234
    Johnson, 234
    Cain's reed-bent, 54
    mountain oat, 94
graybacks, 56, **57**
Great Smoky Mountains Association, 198, 243
Great Smoky Mountains Conservation Association, 242–43
Great Smoky Mountains Institute at Tremont 140–41, 210, 240
Great Smoky Mountains Natural History Association, 243
Green Mountain, 8
Greenbrier Pinnacle, 88, 181
Greenbrier, 32, 147, 151, 169, 177
greenhouse gases, 236
Gum Swamp, 102, 114, 120
Guyot, Arnold, 22

H

Hannah, Evan, 34
Hantavirus, 152
    Newfound Gap virus, 152
    Sin Nombre strain, 152
Happy Valley, 21

Harmon, Mark, 193
Hartmann, Larry, 194
Harvey, Michael, 165
Hazel Creek, 147, 177
Hemlock Inn, Bryson City, North Carolina, 190–91
hemlock woolly adelgid (*Adeleges tsugae*), 81, 86, 225–29, 239
Hemlock Woolly Adelgid (HWA) Control Project, 227
Hemphill Bald, 92
Hen Wallow Falls, 114
Henderson, W. C., 148
Hermann, Ray, 192
Hesler, H. R., 186
Hooper Bald, 230
Horsetail, 53
Hubbs, Carl, 186
Huggins Hell, 89
Hughes Ridge, 62
Hyde, Erin, 111

I

Indian Camp Creek, 70
Indian Flats Prong, 110
Indian Gap, 31, 167
insects, 19, 99–102
    Coleoptera (beetles), 99
        acorn weevil (*Curculio* sp.), 80
        burying beetle (*Nicrophorus tomentosus*), 99
        firefly beetle (*Photinus carolinus*), 99
        ladybird beetle, 228, 229
        Laricobius nigrinus, 228
        Sasajiscymus tsugae, 228
        southern pine beetle (*Dendroctonus frontalis*), 212–**13**
        water pennies, 103
    Diptera (flies), 103
        blackfly, 99, 103
        caddisfly, 103
        cranefly, 103, 104
        dobsonfly, 103
        European mountain ash sawfly, 230
        flies, 230

midges, 99
stonefly, 103
Homoptera (adelgids, leafhoppers), 103
　Adelges piceae, 214
　Adelges tsugae, 225
　balsam woolly adelgid, 65, **66**, 116, **214**–16
　beech bark scale insect, 224–25
　hemlock woolly adelgid, 81, 86, 225–229, 239
　leafhoppers, 219
Hymenoptera (bees, wasps, ants), 99
　bees, 73, 99
　fire ants (*Solenopsis* sp.), 230
Lepidoptera (butterflies and moths), 99, 100–02
　acorn moth, 80
　European gypsy moth (*Lymantria dispar*), **222**–**24**
　hawk-moth, 73
　monarch butterfly (*Danaus plexippus*), 29, **100**–102, 186
　moths, 99
　skippers, 99
　tawny crescent butterfly (*Phycoides batesii maconensis*) 204
Orthoptera (grasshoppers, roaches), walkingstick, **98**
　woodroach (*Crytocercus punctulatus*), 99
Thysanoptera (thrips)
　pear thrips, 229
Trichoptera (caddisflies), 104
invertebrates, aquatic, 102–4
　crayfish, 103
Iron Mountains, 51

J

J. J. English Company, 37
Jenkins, Mike, 194
Jennison, H. M., 36

K

Keohane, Nancy, 101
Kephart, Horace, 59
King, Willis, 37, 176

Komarek, E. V., 36
Komarek, Roy, 36
Kunze, Michael, 194

L

Lake Chilhowie, 111
Langdon, Keith, 183, 194
Larson, Gary, 193
Laurel Creek, 37
Laurel Falls, 188
LeConte Creek, 109–10
LeConte Lodge, 24
Leptospirosis, 233
lichens, 19, 68–69, 203
　number of in park, 55
　British soldier, 68, **69**
　Cladonia cristatella, 68
　Leioderma cherokeense, 203
　reindeer, 90
　rock gnome, 55, 161, 162
　scarlet-crested cladonia. *See* lichens: British soldier
Lindsay, Mary, 193
Little Greenbrier, 43
Little Pigeon River, 11, 107, 115
Little River, **22**, 107, 115, 176, 177, 219
Little River Lumber Company, 37, 42, 49
Little River Trail, 157
Little Tennessee River, 11, 31, 107, 172, 174, 175
lizards, 122, 129
　anole, **123**–24, 239
　broad-headed skink, 124
　coal skink, 124
　eastern slender glass lizard, 123
　five-lined skink, **122**, 124
　northern fence lizard (*Sceloporus*), 123, 185
　southeastern five-lined skink, 124
localities referenced in text (Appendix 1), 245–52
logging, 176
Look Rock, 210, 211
　parking area, 230
　tower, 29
Love, Jason, 101

Low Gap, 27
Low Gap Trail, 27
Lower Cammerer Trail, 128

M

Maddron Bald, 88
Magnesium, 221, 233
Maloney Point, 16
mammals
   species of in park, 141–45
   bats, 160
      big brown, 160
      eastern pipistrelle, 160
      evening, 198
      hoary, 160
      Indiana, 162, 165
      Rafinesque's big-eared, **159**
      red, 160
      silver-haired, 160
      small-footed myotis, 160
      echolocation, 160
   black bear, 29, 35, 36, 38, 40, 80, 81, 151, 153–59, 155, 157, 168, 178, 180, 233, 235
   bison, 33
   cats
      bobcat, **169**, 233
      eastern cougar, 35, 162, 167–69, **168**
      mountain lion. *See* eastern cougar
      panther. *See* eastern cougar
      puma. *See* eastern cougar
      wildcat, 35
   cattle, 34
   coyote, 94, **146**–47, 168, 169, 179, 233
   eastern cottontail, 40
   elk, 29, 33, 34, 92, 167, 177–79, **178**
   European wild boar, 193, 214, 230. *See also* wild hog
   fox, 35, 36
      gray, **142**, 146, 168
      red, 35, 146, 168, 235
   mink, 36, **144**
   mole, 129, **153**–54
      star-nosed, 153–54
   northern river otter, 33, 103, **176**–77
   opossum, 36, **142**, 169, 186

pig, 34
rabbit, 36, 129, 180, 235
raccoon, 36, **144**, 169, 180
rodents
   beaver, 33, 103, 147
   chipmunk, 29, 38, 80, 129, **145**, 150, 186
   eastern woodrat, 150
   golden mouse, 40, 150–51
   jumping mice, 29; meadow, 150; woodland (*Napaeozapus*), 129, 150
   Microtus pinetorum, 129
   muskrat, 36
   Peromyscus, 38, 95, 129; deermouse, 38, 95; white-footed mouse, 95
   red-backed vole (*Clethrionomys gapperi*) 57, 95, 129, 232
   rock vole (*Microtus chrotorrhinus*), 57, 129
   squirrels, 36, 38, 80, 129; eastern gray, 151, 169; fox, 151; northern flying, 57, 165–67, **166**, 216, 239; red, **145**, 151; southern flying, 129, 152
   woodchuck, 29, 40, **150**, 180, 235
sheep, 34, **92**
shrews, 129, 152–53
   northern water, 57, 95, 152, 232
   pigmy, 152
   short-tailed, 152–**53**
   smoky, 57
   Sorex fumeus, 57
   Sorex palustris, 57
   southeastern, 40
skunk, 169
   spotted, **143**
weasels, 129
white-tailed deer, 29, 33, 80, 147–**49**, **148**, 169, 180
wild hog, 80, 168, 230, **231**, **232**, 233
wolf, 35
   gray, 162
   red, 162, 167, 179–81, **180**
Mannis Branch, 110
Maple Sugar Gap, 70
maps
   bedrock geology of the park, **12**

distribution of glaciers, **56**
disturbance history, **40**
Great Smoky Mountains National Park, **3**
movement of the continents, **10**
vegetation of the park, **60**
Maryville, Tennessee, 49
mast, hard, 80, 156, 232
    acorns, 80
    beech nuts, 80
    hickory nuts, 80
    walnuts, 80
Mayor, Adriean, 194
McClure, G. W., 37
Mercury, 208, 210, 211
Metcalf Bottoms, 224
Michaux, Andre, 36, 183
Microhexura montivaga, 57, 162–**63**
Middle Prong of Little River, 37
Middle Prong, Little Pigeon River, 177
Miller Cove, 17
millipedes, 97–**98**, 99
Mingus Mill, 13
Mingus, John J., 33
Monarch Watch, 101
Moore, Steve, 195
mosses, 53
Mount Buckley, 36, 163
Mount Chapman, 63, 163
Mount Collins, 163
Mount Craig, 22
Mount Guyot, 22, 62, 63, 215
Mount LeConte, 14, 16, **19**, 22, 24, 27, 37, 41, 47, 62, 63, 89, 90, 163, 188, 215
Mount Love, 163
Mount Mitchell, 22, 166
Mount Sterling, 36, 63, 215
Mount Sterling Gap, 159
Myrtle Point, 89

N

National Park Service mandate, **47**
natural history collection, 241–42
Necker, W. L., 37
Nelson, Diane, 203
Neuman, R. B., 187

New River, 109
Newfound Gap, 31, 46, 47, 48, 61, 166, 227
Newport, Tennessee, 179
Nichols, Becky, 104, 194
Nickerson, Max, 115
Noland Creek, 177
Noland Divide, 210
Noland, William, 34

O

oak apples, 79
Oconaluftee, 13, 179
    fault, 17
    River, 107, 115, 177
Old Black, 63
Oliver, John, 33, 35
Oliver, Lurany F., 33, 92
Oliver, Mrs. John. *See* Lurany F. Oliver
orogenies
    Acadian, 9
    Alleghenian, 10
    Appalachian, 10, 18, 21
    Taconic, 9
ozone, 65, 208, 209, **210**, 211

P

panther, 167. *See also* cats: eastern cougar
park headquarters, 159
Parker, Chuck, 194, 195
parks as classrooms, 242
Parson Bald, 91
Peine, John, 193
Pelton, Janice, 101
Pelton, Michael, 156, 193
Pepoon, H. S., 36
Peregrine Peak, 20
Peregrine Ridge, 182
Phosphorus, 233
Pickering, John, 194
Pigeon River Valley, 14
Pigeon River, 107
Pilkey Creek, 109–10, 177
Pinnacle Creek, 147
Pittillo, Dan, 193
Pollock, Kenneth, 111
Porcine parvovirus, 233

Potassium, 221
Proctor, 37
Prysby, Michelle, 101
Pseudorabies, 233
puma, 167. *See also* eastern cougar
Purchase Knob, 101, 141, 210, 216
Purchase Knob Appalachian Highlands Learning Center, 211
Pyle, Charlotte, 193

Q

Qualla Boundary, 179

R

Rainbow Cave, 18
Rainbow Falls Trail, 227
Rakes, Patrick, 174
Ravensford Tract, 48, 151, 203
reintroductions, 171–82
    barn owl, 182
    duskytail darter, 175–76
    elk, 177–79
    feregrine falcon, 181–82
    red wolf, 179–81
    river otter, 176–77
    smoky madtom, 162, 164, 172–75
    spotfin chub, 175
    yellowfin madtom, 172–75
Remus, Charles, 146
Renfro, James, 193, 195
reptiles
    species of in park, 121–22, 124–5
    See also lizards; snakes; turtles
Riddle, Jess, 84, 86
Ridge and Valley Province, 8
Road to Nowhere, 191
Roaring Fork, 77
rock types/formations
    Anakeesta formation, 14, 16
    argilite, 16
    arkose, 16
    basement complex, 16
    dolomite, 16
    Elkmont sandstone, 14, 16
    feldspar, 14
    gneiss, 13, 16
    granite, 13, 14
    igneous, 11
    karst, 17
    limestone, 8, 13, 15, 16, 17, 19
    metamorphic, 11
    metasandstone, 14
    metasiltstone, 14, 16
    Ocoee Supergroup, 14, 16, 17
        Great Smoky Group, 14, 16
        Snowbird Group, 14, 16
        Walden Creek Group, 14, 16
    phyllite, 14, 16, 19
    quartzite, 13, 14, 16, 19
    sandstone, 8, 13, 16, 19
    schist, 14, 16
    sedimentary, 11, 13
    shale, 8, 13, 16, 19
    siltstone, 8, 13, 16, 19
    slate, 13, 14, 16, 19
    thunderhead sandstone, 14, **15**, 16
Rock, Janet, 194
Rockefeller, J. D., Jr., 46
Rockefeller, Laura Spelman, 46
Rocky Mountains, 11
Roosevelt, President Franklin D., **48**, 49
Russell, N. H., 53

S

Sag Branch, 71
salamanders, 129, 232
    species of in park, 112–13
    Ambystoma, 115
    Ambystomatidae, 114
    Aneides aeneus, 111
    Desmognathus imitator, 111
    Desmognathus ocoee, 111
    Desmognathus wrighti, 111
    dusky, 114
    eastern red-spotted newt, 116–17, 120
    Eurycea guttolineata, 36
    green, 37, 111
    hellbender, 103, 112, 115
    Jordan's (red-cheeked), 115–**16**
    imitator, 115–16
    long-tailed, 113
    lungless, 97, 115

marbled, 112, 114
mole, 114, 198
pigmy, 37, 113, 116, 216
Plethodon glutinosus, 111
Plethodon jordani, 36, 111
Plethodon oconaluftee, 111
Plethodontidae, 97, 115
red eft, 116–17
red-cheeked salamander, 115, **116**, 232
spotted, 29, 112, 114, 120
three-lined salamander, 36
Sams Creek, 110
Savage, Tom, 129
Schoolhouse Gap, 37
Sequoyah, 33
Sharp, Jack, 187
Shenandoah National Park, 226
Shields A. Randolph, 33, 40
shrubs
    number of in park, 53
    American mistletoe, 74
    blackberry, 64, 67, 81, **210**
    bush-honeysuckle, 64, 234
    dog-hobble, 67–**68**, 81, 227
    eastern leatherwood, 204
    flame azalea, 24, 29, 79, **93**
    hobblebush, 64, 81
    huckleberry, 79
    hydrangea, 67
    mountain laurel, 24, 29, 59, 81, 85, 88, **89**, 90
    rhododendron, **74**, 88, 140
        Carolina, 64, 89
        Catawba, 29, 64, **67**, 81, 85, 89
        Rhododendron maximum, 140
        rosebay, 29, 59, 64, 67, 73–74, 79, 81, 89, 221, 239
    sandmyrtle, 89, **90**
    spicebush, 29, 74
    Virginia spiraea, 55, 162
    wild raisin, 64
Shute Peggy, 174
Shute, J. R., 174
Sierra Madre Mountains, 101
Sierra Nevada Mountains, 11
Simons, Ted, 111, 137, 140, 226

Skunk Ridge, 165
Smokemont, 15, 31, 37, **42**, 151
Smoky Mountain Field School, 242
snails, 7, 19, 98–**99**, 100
snakes, 124–25, 129
    species of in park, 124–25
    adaptations for feeding, 125
    black rat snake, 126, 127
    blacksnake, 186
    Colubridae, 125
    copperhead, 126, 127, **128**–29
        facial pit, **128**
    corn snake, 125, **126**
    eastern garter snake, **125**
    eastern kingsnake, 125, 126
    northern water snake, 126, **127**, 186
    pit vipers, 125
    snake skull, **126**
    timber rattlesnake, 126, 127, **128**–29
    Viperidae, 125
soils, 19, 20
Southern Appalachians, 19
Sparks, Tom, 167
spruce-fir moss spider, 57, **163**–64, 239
Stony Branch, 226
Stupka, Arthur, 37, 51, 59, 101, 182, 183, **184**–91, **189**, 239, 241
Stupka, Margaret, 190
Sugar Orchard Branch, 70
Sugarland Mountain, 70
Sugarlands, 14, **26**, 70, 158, 227
    visitor center, 227
Super, Paul, 101
Sutton Ridge, 128
Sweatheifer Trail, 167

T

Tabcat, 177
Tallassee, 32
Tardigrade, **202**–3
Tennessee River, 21
Tennessee Valley, 19
Tonsberg, Tor, 203
Townsend, Col. W. B., 42
Townsend, Tennessee, 157
Toxoplasmosis, 233

Trail of Tears, 33
trees
    number of in park, 54
    autumn color change, 26–27
    Allegheny serviceberry, 62, 66, 79, 85
    American beech, 62, 66, 67, 69, 81, 83, 229
    American chestnut, **38**, 69, 70, 79, 216–19, **217**, **218**, 220
    American holly, 81
    American mountain-ash, 62, 63, 230
    American sycamore, 83
    ash, 69, 83, 229
    basswood, 69, 229
    birch , 147, 229
        river, 147
        yellow, **73**
        sweet, 81
    black gum, 74, 79, 86
    black locust, 79, 83
    blueberry, 64
    buffalo nut, 29
    bush-honeysuckle, 64, 234
    butternut, 229
    Carolina silverbell, 85
    cherry
        black, 62, 66, 69, 81, 85, 209, 210, 229
        pin, 29, 62, 63, 66, 81, 85
    cinnamon clethra, 81, 85
    common alder, 59
    Crateagus iracunda, 59
    cucumbertree, 71, 83
    devil's walkingstick, 59, 85
    dogwood, **220**
        alternate-leafed, 59
        flowering, 26, 29, **53**, 86, 147, 185, 186, 219
    Fraser fir, 61, **62**, 63, 65, 66, 163, 214, 215, 216
    hawthorn, 59
    hemlock
        Carolina, 226
        eastern, 66, 69, 81, 83, 84–85, 86, 87, 140, 220, 221, 222, 225, 226, 227, 229
    hickory, 26, 83, 235
        pignut, 83
        shagbark, 69, 83
    hop-hornbeam, 69
    ironwood, 69, 71
    locust, 235
    magnolia,
        fraser, 29, 69, 71, 72, 85
        umbrella, **72**
    maple, 229, 235
        mountain, 62, 63, 66
        red, 29, 66, 69, 81, 83, 85, 86, 185, 221
        striped, **66**, **67**, 69, 85
        sugar, 66, 69, 70, 81, 83, 229
    mimosa, 234
    mountain stewartia, 59
    oak, 87, 235
        black, 79
        chestnut, 79
        northern red, 83
        scarlet, 26, 74, 79, 87
        white, 79, 83
    persimmon, 26
    pine, 86, 87, 212, 235
        pitch, 79
        shortleaf, 79, 165
        table-mountain, 79, 87, 213
        Virginia, 79, 86, 165
        white, 79, 83, 86
    red elderberry 64
    round-leaved currant, 64
    sassafras, 79, 80–81, 209, 235
    scarlet elder, 81
    silverbell, 29, 69, 81
    sourwood, 26, 29, 79
    spruce
        Norway, 234
        red, 61, **62**, 63, 66, 83, 85, 209, 212
    stolon-bearing hawthorn, 59
    sumac, 26, 185
        shining (winged), 59
        staghorn, 59
    sweetgum, 83, 209
    sycamore, 147
    tree of heaven, 234

Tsuga canadensis, 140, 226
Tsuga caroliniana, 226
tuliptree, 26, 29, **69**, 70, 81, 83, 147, 185, 209, 210
umbrella magnolia, 29, 71, 72
Viburnum sp., 59
white basswood, 83
white poplar, 234
witch-hazel, **72**
yellow birch, 29, 63, 66, 69, 73, 81
yellow buckeye, 62, 66, 69, 83, 85
yellow-poplar. *See* tuliptree
Tremont, 37, 49, 127, 138, 140, 180, 193
Trillium Gap Trail, 227
Tuckaleechee Cove, 17
turtles, 121
    common musk turtle, 198
    common snapping turtle, 122
    Cumberland slider, 198
    eastern box turtle, **121**, 122
Twentymile, 177
Twin Creeks, 32, 47, 101, 193, 194, 210
Twin Creeks Natural Resources Center, 194
Twin Creeks Science Center, 183, 195, 240, 241–42

U

Unaka Mountains, 11, 13, 51
Uplands Field Research Laboratory, 192, 193, 194

V

vines, 53
    number of in park, 53
    blackberry, 209
        ozone damage of, 210
    climbing euonymus, 234
    Dutchman's-pipe vine, 77
    English ivy, 234
    poison ivy, 236
Voorhis, Ken, 240

W

Walker, Louisa, 44
Walker Prong, 166
Walker sisters, **43**–46
Walker, Dan 43
Walker, Felix, 33
Walker, John, 43
Walker, Louisa, 43, **44**, 45, 46
walkingstick, **99**
water quality, 211–12
Waynesville, North Carolina, 157
Wear Cove, 17
Webster, Chris, 58
Weigl, Peter, 166
Welch Branch, 231
Weller, W. H., 37
Wells, B. W., 53
West Prong, Little Pigeon River, 166, 177
White, Peter, 193, 195
White Oak, 179
Whiteoak Sink, 18
Whittaker, R. H., 187
Whittaker, Robert, 51
Wildflower Pilgrimage, 187, 192
wildflowers
    American woodsorrel, **63**, 64, 67
    bee-balm, 29
    bloodroot, 29
    bluets, 29, 64
    butterfly-weed, 29
    cardinal flower, 27, 29
    chickory, 27
    clematis, 186
    columbine, 29
    common fawnlily, 67
    creeping bluet, 67
    creeping phlox, 29
    crested dwarf iris, 29
    crinkleroot, 67
    cutleaf coneflower, 209, 210
    deptford pink, 24, **25**
    doll's-eyes, 75–**76**
    downy rattlesnake plantain, 82
    dutchman's breeches, 29
    early meadowrue, 29
    fire pink, 24, 29, **54**
    fly poison, 29
    foam flower, 29

wildflowers (cont.)
   fringed phacelia, 67, 68
   galax, 29, 79
   goldenrod, 29, 185
   great starwort, 67
   heart-leaf aster, 29
   heart's-a-bustin, 74
   indian-pipe, 64, **75**
   jack-in-the-pulpit, **76**–77
   Joe Pye weed, 29
   ladies' tresses, 29
   large-flowered bellwort, 29
   little brown jug, 77
   mercury spurge, 204
   milkweed, 100–102, 209, 210
   mullein, 234
   multiflora rose, 234
   orchids, 53
      purple-fringed, 29
      yellow-fringed, 24, 29, 53
   pallid violet, 64
   partridge-berry, 81–**82**
   pink lady's slipper, **25**
   pink turtlehead, 29, **64**
   pipsissewa, 79
   pussytoes, 79
   Rugel's Indian plantain. *See* Rugel's ragwort
   Rugel's ragwort, 29, 36, 54
   serrinated skullcap, 204
   sharp-lobed helpatica, 29
   showy orchis, **25**, 29
   speckled wood lily, 29
   spreading avens, 55, 161–62
   spring beauty, 29, 64, 67, 232
   touch-me-not, 29
   trailing arbutus, 29, **79**
   trillium, 149
      painted, **64**
      wake robin, 29
      white, 29
      yellow, 29
   trout lily, 29, **75**
   turk's cap lily, 29, **64**
   white baneberry. *See* doll's-eyes
   white fringed phacelia, 29, **68**
   white wood aster, 64
   wild geranium, 29
   wild ginger, 77
   wild hydrangea, 81
   wintergreen, 79
   yellow bead lily, 64, 67
   yellow lady's slipper, **25**, 29
   yellow star grass, 29
   yellow trillium, 29
   yellow-fringed orchid, 24, 29, **53**
wildfire, 221
Wildflower Pilgrimage, 23
Winding Stair Branch, 110
Wood, Matthew, 218

Y

Young, Rob, 91

Z

Zank, Ben, 194
Zeigler, William, 92
Zinc, 233